THE MATTER MAN AND THE THEORY OF ADAPTABILITY

Modern Scientific Theory on Discrimination

SECOND EDITION

DESMOND AYIM-ABOAGYE

THE MATTER MAN AND THE THEORY OF ADAPTABILITY

Modern Scientific Theory on Discrimination

SECOND EDITION

By

Desmond Ayim-Aboagye
Ph.D., Associate Professor
Åbo Akademi University, Finland
Uppsala University, Sweden

Lulu.com, North Carolina, USA

Copyright © 2006, 2007 by Lulu.com

ISBN 978-1-84728-137-1

All rights reserved.

Reproduction or translation of any part of
this work beyond that permitted by Sections
107 and 108 of the 1976 United States Copyright
Act without the permission of the copyright
owner is unlawful. Requests for permission or
further information should be addressed to Permissions Department, Lulu.com

"For, though these properties also were naturally inherent in the figures all along, yet they were in fact unknown to all the many able geometers who lived before Eudoxus, and had not been observed by any one. Now, however, it will be open to those who possess the requisite ability to examine these discoveries of mine. They ought to have been published while Conon was still alive, for I should conceive that he would best have been able to grasp them and to pronounce upon them the appropriate verdict; ...."

Archimedes (In his fourth book, *On the Sphere and Cylinder*, Book One)

TABLE OF CONTENTS

BOX 10

FIGURE 10

PREFACE TO THE READER 11

CHAPTER 1: TOWARD A THEORY OF NON-DISCRIMINATION 15
- PREAMBLE 15
- THE PROBLEM 16
- DEFINITIONS 16
  - *Definition 1* 16
  - *Definition 2* 17
  - *Definition 3* 18
  - *Definition 4* 18
  - *Definition 5* 19
  - *Definition 6* 19
  - *Definition 7* 19
  - *Definition 8* 19
  - *Definition 9* 19
  - *Definition 10* 20
- POSTULATES 20
  - *Postulate 1* 20
  - *Postulate 2* 20
  - *Postulate 3* 21
- SCHOLIUM 21

CHAPTER 2: THE PRINCIPLE OF NON-DISCRIMINATION LAWS, OR AXIOMS AND OTHER PROPOSITIONS 24
- PROPOSITION 1 24
- PROPOSITION 2 24
- LAW 1 24
- LAW 2 25
  - *Corollary 1* 25
  - *Corollary 2* 25
  - *Corollary 3* 26
  - *Corollary 4* 26
- AWARENESS AND FEAR 26

Flight can be an exploratory strategy ... 27
Device for relieving fear ... 28
Perceptual curiosity, learning and conditioning ... 29
Vital notes on human behaviour and conditioning ... 29
Scholium ... 31
Impetus Principle ... 33
Induction Theorem ... 34
Omega Theorem of the Principle of Non-discrimination ... 34
Lucifer Theorem ... 35
Sola Fide Theorem ... 35
Scholium ... 36

CHAPTER 3: THE SOFT CONDITIONING INCENTIVE ... 37

Of the Principle of Common Safety ... 38
The *Subjectus* and "Union" Contract ... 40
Some Perceptions and Actions of those who Discriminates ... 41
Scholium ... 42

CHAPTER 4: THE HARD CONDITIONING INCENTIVE ... 44

The Dominant principle ... 46
The Sovereign: the puppet and the forfeited ... 46

CHAPTER 5: OBJECTIONS TO THE UPRISING OF RACISM ... 48

The Three Predictions of Racism ... 48
Racism Causes Harm to Oneself ... *48*
Racism Causes Harm to the Physical World ... *48*
Racism Causes Harm to the Feelings and Integrity of Other Races ... *49*
The Objections ... 49
Scholium ... 51

CHAPTER 6: THE THEORY OF SUPERIORITY COMPLEX: PERSONAL AND COLLECTIVE EXPERIENCE ... 52

Equilibrium Concept ... 52
Equilibrium, Democratic Tradition and Complex ... 54
Complex and Equilibrium ... 56
Linearity and Non-linearity of Complex ... 57
Pinnacleconductivity and superfluity ... *59*
The Six Important Formal Axioms or Principles of Superiority Complex ... 60
Axiom 1 ... *60*
Axiom 2 ... *61*
Axiom 3 ... *61*
Axiom 4 ... *61*
Axiom 5 ... *62*
Axiom 6 ... *62*
Complex and the Principle of Self-sufficiency ... 63
Complex and Recognition ... 63
Complex and the Selfishness Theory ... 63
Consensus ... 65
Closed and Open Groups ... *65*
Scholium ... 66
Superiority Complex can be transmitted ... 67
The Conspiracy Theorem ... *67*
Minutron theory ... *67*
Minor Theorems ... 70
Minutrons Perforation ... 73
Scholium ... 74
Miscellaneous ... 75

CHAPTER 7: DISCRIMINATION AND THE THEORY OF ALLIANCES 80

- THE THEORY OF ALLIANCE 80
- THE NECESSITY OF ALLIANCES 80
- DIFFERENT ALLIANCES 81
  - *1. Professional Alliances* *81*
  - *2. Marriage Alliances* *82*
  - *3. Underground Alliances (Underground groups)* *82*
  - *4. Sovereign Alliances* *82*
  - *6. Voluntarius Alliances* *83*
  - *7. Multiple Alliances* *83*
- THE PROBLEM OF NEUTRALITY 83
- THE PROCESS OF DEFENCE, GETTING HELP OR BARGAINING FOR AN ALLIANCE PARTNER 84
- EMPIRICAL EVIDENCE: THE GROWTH OF ALLIANCES 84
  - *Alliance as a General Form of Protection* *84*
- SCHOLIUM 85
- DEFINITIONS 85
  - *Definition 1* *85*
  - *Definition 2* *86*
  - *Definition 3* *86*
- AXIOM OR PROPOSITIONS OF ALLIANCES 86
  - *Axiom 1* *86*
  - *Proposition 1* *86*
  - *Proposition 2* *86*
  - *Theorem 2* *86*
  - *Proposition 3* *87*
  - *Theorem 3* *87*
  - *Proposition 4* *87*
  - *Proposition 5* *87*
  - *Proposition 6* *87*
- MAJOR AND MINOR THEOREMS 87
- SCHOLIUM 90

CHAPTER 8: OF CLIMATIC LAW AND HUMAN BEHAVIOUR: NATURE AND ITS PECULIARITIES 91

- THE THEORY OF SUB-RACES: FABLES 91
- THE THEORY OF ADAPTABILITY 93
  - *AXIOMS OF THE THEORY OF ADAPTABILITY* *100*
    - Axiom 1 100
    - Axiom 2 100
    - Axiom 3 100
    - Axiom 4 100
    - Axiom 5 100
    - Axiom 6 101
    - Axiom 7 101
  - *Scholium* *101*
- SOME IMPLICATIONS OF THE THEORY OF ADAPTABILITY 101
- IMITATION AND THE THEORY OF ADAPTABILITY 103
  - *The Primary Frontis* *104*
  - *The Secondary Frontis* *105*
  - *The Main Frontis* *105*
- SAPIENCE AS A MYTH 106
- LANGUAGE, DEVELOPMENT AND CONTINUITY 107
- IMITATION, TALENTS AND ADAPTATION 108
  - *The Processes* *108*
- "CRISES SITUATIONS" AND "AGGRESSIVE BEHAVIOUR" AS NECESSARY FAMILIAR NOTIONS IN ADAPTATION 109
- ADAPTABILITY AND THE THEORY OF SCIENCE 111
- ADAPTABILITY, SCIENCE METHODS, AND SELF-EFFICACY 113
- IMITATION AND SCIENTIFIC CONTRIBUTION 114

SCHOLIUM 116
MAJOR AND MINOR THEOREMS OF THE THEORY OF ADAPTABILITY 117
MISCELLANEOUS 120
SCHOLIUM 124

CHAPTER 9: OF SYSTEMS OF POLITICAL DISCRIMINOLOGY 125

OF THE PRINCIPLE OF RESTRAINTS 125
OF ALLOWANCE OF THE *SUBJECTUSES* 125
OF UNFAIRNESS OF RESTRAINT MEASURES 126
ON THE QUESTION OF REVENUE AND THE *SUBJECTUSES* 126
THE LEGISLATOR AND STATESMAN 128
SCHOLIUM 129

GENERAL SCHOLIUM 131

APPENDIX A 133

PERSPECTIVES IN RELATION TO LEARNING AND THE THEORY OF BLOOD LANGUAGE 133

PREAMBLE 133
   The Purpose *134*
DIFFERENT PERSPECTIVES ON LEARNING 134
THE COGNITIVE CONSTRUCTIVIST APPROACH 135
   Jean Piaget and his followers *135*
   Problems: classroom learning in the constructivist perspective *136*
SOCIOCULTURAL AND THE INTERACTIONISTIC PERSPECTIVE 137
   The Contributions of Vygotsky *138*
   Situated learning theory *140*
   Classroom in the sociocultural theorists' view *140*
   Roger Säljö and the concept "practice" *141*
      Practice in school 141
      Three Aspects of Practice 142
THE LANGUAGE FOCUSED APPROACH 143
   Sociolinguistics: "talking" knowledge *143*
   Language as cultural artefact *144*
   Linguistic genre as significance in cognition *144*
PHENOMENOGRAPHY AND VARIATIONS THEORY 145
   Toward experiential description of learning *146*
   The learner and his learning *146*
   Variation in learning *147*
   Three Aspects of learning: Variation, Discernment and Simultaneity *147*
THE THEORY OF BLOOD LANGUAGE 149
DEFINITIONS 150
   Definition 1 .... *150*
   Definition 2 .... *150*
   Definition 3 .... *150*
   Definition 4 .... *151*
   Definition 5 .... *151*
   Definition 6 .... *151*
   Definition 7 .... *151*
   Definition 8 .... *151*
AXIOM AND PROPOSITIONS 151
   Axiom 1 .... *151*
   Axiom 2 .... *152*
   Axiom 3 .... *152*
   Comprehension—Proposition 1 .... *152*
   Theorem 1 .... *152*
   Description—Proposition 2 .... *152*

Theorem 2	*152*
Originality—Proposition 3	*153*
Theorem 3	*153*
Accuracy and Sharpness—Proposition 4	*153*
Theorem 4	*153*
Carrying out responsibilities—Proposition 5	*154*
Theorem 5	*154*
MAJOR AND MINOR THEOREMS	154
Definition 9	*157*
Axiom 4	*157*
SCHOLIUM	160
GENERAL SCHOLIUM	160

APPENDIX B 162

OF DARWINISM AND THE DEVELOPMENT OF SUPERIORITY COMPLEX PERSONALITY-TYPE 162

INTRODUCTION	162
THE PURPOSE	162
BASIC SCIENTIFIC IDEAS PRECEDING CHARLES DARWIN	163
THE CENTRAL ELEMENTS OF DARWIN'S THEORY	164
SOCIAL DARWINISM: THE FOUR PHASES	165
CRITICISMS FROM MODERN PSYCHIATRY POINT OF VIEW	166
DARWINISM AND THE PSYCHO- S COMPLEX PERSONALITY-TYPE	167
Psycho-S Complex: Signs or Symptoms	*168*
Delusion	168
Auditory hallucination	169
Prejudice predisposition	169
Manipulative	169
Winmust	170
Superexaggerare	170
Superwunsken	170
Infantile notions	170
Hannibal Odyssey	171
SCHOLIUM	171

LIST OF THE PRINCIPAL AND MINOR WORKS CONSULTED. 173

INDEX 179

Box

Box 1: Formulae of the principle of non-discrimination law. 31
Box 2: The common safety principle 38
Box 3: The dominant principle 44
Box 4: Adaptable behaviour toward development 111
Box 5: The Talented 115
Box 6: Talents for nation building 115
Box 7 Formulae for successful development leading to adaptability 123

Figure

Figure 1. Less fear correlates with higher employment rate 37
Figure 2. Less fear correlates with less hate 37
Figure 3. Union Contract Marriage (We live as friends in marriage when there is no more love) 41
Figure 4. Greater fear correlates with low employment rate 45
Figure 5. Greater fear correlates with greater hate 45
Figure 6. Linear graph for the variables complex and age 59
Figure 7. Linear graph for variables complex and age (non-linear) 59

Preface to the Reader

Since the inception of the old colonial era many studies have been devoted into the investigation of the relationship between different races. During those pioneering days the general purpose of research was to comprehend the distinctive characteristics that differentiate one race from another. A corollary aim was to find an appropriate manner where the different races could live harmoniously with one another under the colonial regimes, and subsequently under the umbrella of the imperialist regimes. These were the periods where the powers of imperialist regimes and the colonial states were not being challenged, and the climate for research was generally excellent. These researches would also help the colonial masters to become aware of the needs of each people or races, so that they could easily attend to them. These researches were later to become the centres of contentions among the different scholars in the fields of anthropology, ethnology, sociology, history, cross-cultural psychology, transcultural psychiatry, and many others. Scholars from these different disciplines had their own way of carrying out their investigations; while others had their eyes focused on describing objectively what they saw as distinctive characteristics of a particular race, others became engrossed with what they felt was peculiar within certain races that make them less human as compared with themselves.

Any one who had followed the reductionism that took place in the colonial days would discover a series of blunders that these pioneer scholars made. It was as if in their histories or historical backgrounds everything was already made and sent down from an outer planet into the earth. In their attempt to describe how people live in these different cultures they forgot to review their own past, which was filled with nakedness, poverty, slavery, brute wars that were barbaric in nature, dehumanising ways of treating people, and their own greedy nature that had preceded the enlightenment and time immemorial. The Roman conquerors found out that many nations were immersed in these miserable conditions before they began to transform them and their brute cultures. Others could not be allowed to use their own blood language or write it out, which had made them disillusioned and ignorant as a result of being compelled to use another race or people's language (e.g., Latin). A priest could in those days ask an individual to pay some indulgence because this could easily transform the dead uncle from his own condition of torture to that of relief and a peaceful rest. The seed of religion blinded men's thoughts and made them ignorant, like any other race on earth. Thanks to easy contacts that occurred between different people in the then-known world, people were able to learn easily how to clothe themselves well and build good houses to protect them from becoming extinct from the harsh environmental conditions of the planet earth. This pre-eminence of erecting weather-adaptable houses to protect them, sometimes erecting a higher building foundation to escape the rising snow, gave the idea of tall houses or storeys. The good clothing to protect the naked body was due to the extremely harsh weather conditions.

Any-one who will read this treatise is cautioned to be more objective in his thinking and to decide for himself or herself what are the implications. Discrimination, which originates because of a particular race feeling that they originate from a superior species that I consider as an *illusion*, will rightly, be abandoned because this myth has no basis. It was certainly created, and like all inventions that have become accepted as part of a tradition, it has never been questioned. It is something that only people in delusion can glorify themselves as existing in their private sick world. The style of this treatise that follows the geometrical fashion will enable scientist, to investigate further as the propositions and the deductive approach ensure that scientific investigations can be carried out to confirm or disconfirm those hypotheses stated in the form of axioms or propositions. It is to this end that I consider this treatise as not complete, as a whole life will be spent to polish and make it useful in the combat of discrimination in the twenty-first century.

In Chapter 1, *Toward a Theory of Non-Discrimination,* and Chapter 2, *The Principle Of Non-Discrimination Laws, Or Axioms And Other Propositions*, the principle of non-discrimination law explains that human behaviour in terms of discrimination is associated with fear of the unknown. The axioms of non-discrimination are presented and supported with empirical evidences from well-known behaviour theorists. The Chapters argue that discrimination like all other problems with human beings, has a cause and can be tackled. Tangible proposals as to how discrimination in work places can be combated are presented, especially in Chapter 2. It is firmly established that not only the government of a country but also individuals in the country in question, can contribute in the dismantling of these inappropriate behaviours that originate discrimination in society. The focus is more on the individual rather than any doing of a government of a particular state. As far as discrimination is concerned, the responsibility is far removed from the domain of the government to that of the individuals in the society.

The Soft Conditioning Incentive, which Chapter 3 discusses, concerns how the common safety principle should be carried out in the society that is serious in obtaining the equilibrium level. Here some characteristics of the individual or society that practices discrimination are shortly elaborated and the conclusion being that too much concentration on superiority makes societies to fail in their quest to implement the necessary measures that are vital.

In Chapter 4, *The Hard Conditioning Incentive*, the dominant principle is introduced and defined, and is considered as important for any society that intends to reach the higher equilibrium level. The idea of allowing the *subjectuses* to occupy certain higher positions not only strengthens the power of the sovereign; it demonstrates its legitimacy. By ensuring that these individuals become part of the ruling class the sovereign will do much service to themselves, as well as communicate a positive message to the entire subjects of every society.

The ancients were not aware and probably had no sound knowledge of discrimination, since there was another evil object that occupied their minds in those days—slavery. Since we are concerned with the manner of sorting out people that can be considered as a negative behaviour, a separate chapter is devoted to objections to the employment of racist ideologies as a vehicle for supporting discrimination. These objections not only point out the necessity of the sovereign doing something to combat discrimination, but also they possess inherent suggestions that a distinct field of study should be created to investigate these evils that beset the modern society. These are made clear in Chapter 5, *Objections to the Uprising of Racism,* of this treatise, which renew the basic problem of the phenomenon of discrimination.

In Chapter 6, *The Theory of Superiority Complex: Personal and Collective Experiences* is proposed. Personal and collective experiences contain certain hidden mechanisms that can work their way into the unconscious mental states of some human organisms, such that those that cannot be shielded can become vulnerable, and as a result resort to crime. The theory of superiority complex was developed as a result of the need that arose to describe why certain individuals or groups view themselves as superior to others. And, as a result, project themselves above all others, and while at their pinnacle in this miserable condition attempt to cause injury or disturbances to others they usually regard as inferior. This chapter for the first time proposes this obsession with superiority by certain individuals as a kind of disorder that consequently causes them to possess abnormal mental states in their illusory private world. The formal axioms of superiority complex are presented for analysis. There is a call for the necessity of putting them to test or verification. It is argued that this abnormal mental status of the psycho-superiority complex individuals cause them to injure or frustrate other peoples' plans in the society in question, that makes it difficult for certain innocent citizens to actualise their potentials. This in the end prevents the latter from enjoying maximum security and pleasure that they are supposed to enjoy in their host society.

In Chapter 7, *Discrimination and the Theory of Alliances* by means of the theory of alliance, some suggestions are proposed as capable of dismantling or putting discrimination into check. These propositions are not new but, in my opinion, they have never been given serious consideration. They are principles that are deduced from the forces operating as opposition to the integration of the *subjectuses* into the society. For in one way if the forces that are orchestrating these evil behaviours are neutralised, there is no way they will be able to work effectively as a team to disorganise the business of the progressive society. The theory of alliance ensures that certain forces by which the discrimination behaviour are revealed, and the causes not yet properly known, are either impelled to have relations with one another or toward other members of the society. When people cohere in relations, these may help repel the negative ideas or thoughts. Since these forces are unknown or are in hiding due to their cowardice or feeble-minded behaviour, the alliance idea may hinder or aid to let them surface and be properly tackled.

In Chapter 8, *Of Climatic Law and Human Behaviour: Nature and Its Peculiarities,* the theory of adaptability explains how certain laws of nature contribute in a dynamic manner to provide certain basics of man's intelligence. The theory utilises the deductive approach and argues that nature, and not only the human significant others, contribute to the acquisition of intelligence by the human species. Through imitation and adaptation human species have survived and spread all over the planet earth. The theory's greatest contribution is that it attempts to solve the problem of why other species are considered to be fortunate to have primary intelligence, more than others are. The use of the original (blood) language of certain species and the difficulties of certain environments enable others to be more ***aggressive*** in terms of how they ingeniously imitate others and adapt to the planet earth.

From the study of the common safety and the dominant principles, we are capable of deriving the manner the statesman functions in the field of discrimination. That the statesman or legislator should be equipped with puissant knowledge on how to assist and tackle the overall subsistence, both economic aid and employment to the *subjectuses* is imperative. The science of political discriminology makes its utmost aim to caress the *subjectuses*, and to consider them as necessary citizens in the host society. This branch of study proposes to bridge the gap between the *subjectuses* and the native citizens in order to enrich both the sovereign and all ordinary people in the host country. This, which becomes the topic for concluding this treatise, is presented in Chapter 9: *Of Systems of Political Discriminology*

At the Appendix A, *Perspective in Relation to Learning and The Theory of Blood Language,* I discuss the theories advanced on the acquisition of learning as a way of suggesting that the employment of the "language of the blood" for learning in the sciences is vital. I show that since many of the sciences deal with accuracy and detest being in errors, only the utilisation of the blood language for imbibing learning can aid the learner to be free from unnecessary errors. For even in the arts; there are no error, but in the artificers. According to Sir Isaac Newton, "He that works with less accuracy is an imperfect mechanic; and if any could work with perfect accuracy, he would be the most perfect mechanic of all, for the description of right lines and circles, upon which geometry is founded, belongs to mechanics." "Geometry," according to Newton, "does not teach us to draw these lines, but requires them to be drawn, for it requires that the learner should first be taught to describe these accurately before he enters upon geometry, then it shows how by these operations problems are solved."[1] Therefore, a positive recognition and an acceptance of the conceptual frameworks that explain how learners imbibe their learning needs to be reviewed, and to suggest how scholars can fix in this idea of employing the blood language solely to acquire knowledge. We should not ignore the occasional scrutiny of these old perspectives, especially when new insights emerge in the domain of science. To neglect the occasional re-examinations, we may not be able to unravel the intricate problems that still beset man with regard to manners in which the young learners assimilate their learning. This appendix presents an analysis of the perspectives that furnish how learners acquire their overall learning in the context of classroom. The existing theories, as well as new ones are inquired into closely to unveil their similarities, differences, and also contentions that exist among them. The results reveal that; though each perspective contributes uniquely to our understanding, there is still the possibility of blending the perspectives that can be useful to wholeness comprehension of classroom learning. But the employment of these perspectives should recognise the need to imbibe knowledge with the individuals' original language that is fixed in the blood, to ensure accuracy, originality, sharpness and easy imitation and adaptation.

The Appendix B, *Of Darwinism and the Development of Superiority Complex Personality-type,* takes up the issue of the genesis of the development of the Psycho S Complex personality. This chapter can be regarded as a continuation of Chapter 6, where the Psycho S Complex personality is traced back to the work of Charles Darwin concerning his theory of the *Origin of Species.* And more partially to the work of a Swedish Botanist and Scholar, Carl von Linné, who became the first scholar to have employed the terminology "Homo sapiens". This chapter furnishes us with some historical issues, as well as provides us the psychiatry symptoms of those individuals who become engrossed in hostility towards the *subjectuses.* Here, one of the laws of the theory of superiority complex is proved, namely: "Superiority complex is in human beings universally, but its intensity is proportional to the increase gained in knowledge." This chapter gives a strenuous stress and a signification on the proposition, which states that "War is a symptom of a disorder." In fact war is the most primitive method in settling cases or a dispute no matter the sophistication it has become.

As mentioned above, this treatise is a preparatory one to precede a lifetime work that will be devoted in combating the phenomenon of discrimination in this physical world. In asserting this, I do not intend to mean that other researchers have not been working on it, because there is a large body of research that has been accumulated on the topic already. But rather, I personally believe that given better theories concerning the human nature, people will come to appreciate each other much better. This will open the minds of people about the defects of those individuals who become engrossed in this negative behaviour.

I am grateful to Professor Nils Holm, Associate Professor Siv Illman, and Department Secretary Anne Holmberg of Åbo Akademi University for their devotion beyond the call of duty in reading and the preparation of this manuscript. I also want to thank Professor Owe Wikström at the Uppsala University who has inspired me in numerous ways.

I would like to thank the following for permission to use their Library: Uppsala University Library, the Carolina Rediviva and the Department of Economic Science Library, Uppsala, that contains many books on Physics and Astronomy. I also found useful discussions on certain important topics relevant to my work from the Internet Giant, Google. I was greatly inspired by the excellent treatises of Sir Isaac Newton, *The Principia. Mathematical Principles of Natural Philosophy*; Albert Einstein's Relativity Theory and many of his papers written on this theory; Stephen Hawking on his numerous books; Hannes Alfvén about his

---

[1] Motte, 1952: 1.

Plasma theories, Werner Heisenberg's "Uncertainty Principle", and James Clerk Maxwell's Electromagnetic Theory.

Finally, I am particularly grateful to my elder son Yaw Ayim-Aboagye, at Georgetown University, a law student who made constructive criticisms of the final manuscript. He advised me concerning the design of this treatise, as well as encouraged me to publish it.

Desmond Ayim-Aboagye
Associate Professor, Åbo Akademi University
May 30, 2006

Chapter 1: Toward a Theory of Non-Discrimination

Why there is one Body in our System qualified to give Light and Heat to all the rest, I know no Reason, but because the Author of the System thought it convenient; and why there is but one Body of this kind I know no Reason, but because one was sufficient to warm and enlighten all the rest.

Sir Isaac Newton

Preamble

As one move around in Europe and some other Western countries, the word 'discrimination' is not uncommonly heard from the utterances of many migrants who have moved from the less-developed world and certain Third World countries to the former, in search of riches or fortunes. Discrimination is also reported among some of these migrants who have moved from their countries of origin due to wars or persecutions. In the different quarters of these developed countries discrimination is chiefly reported in the area of labour employment. What happens is that though the host country is prepared to receive them into its well-polished society, on the other hand, some authorities and citizens of the land are not willing to freely offer them work or employment that will help them to acclimatise in their new-found milieu. The issues involving discrimination is currently seriously reported in the many different countries, as now and then people of different colour are refused jobs that they believe they have qualifications for. Is it only because there is lack of resources on the side of the host government to create jobs for these migrants that lead to discrimination in these work places? Or is it because the migrants usually have not the required competencies that most of these jobs demand? Is it because there are usually no vacancies for the employers to absorb these migrants? Does the discrimination problem stem from the fact that migrants often lack appropriate qualifications, the required language to carry out responsibilities, valid resident permits that should enable them to work, or are they handicapped in some way that should prevent them to function properly in the work environment? Assuming that some of these problems mentioned above are behind the major reasons why most migrants find themselves discriminated, could there be another tangible reason or reasons why migrants who have even been adopted as citizens of the host country would be discriminated against? Could it be that the problems of discrimination is not at all an issue concerning "who gets what employment," or "what sort of person is to be employed," or "what kind of qualification one possesses," but rather what sort of persons does the migrant worker meet at the work places? In other words, what sort of person does the migrant confront at his work place and what attitude does the host person possesses or reveal? Could there be cultural barriers that bar certain specific individuals to enter into the job market of their host country?

This analysis aims at unveiling appropriate theory that could explain why many migrants who migrate to foreign countries, and even though possess the necessary documents (in some cases have adopted citizenship), still find themselves being discriminated against in the work environment. The approach is not only to provide explanation for the phenomenon itself, but also to clarify the phenomenon as human organism's problem that can be tackled. Furthermore, the investigation goes a bit further to provide some suggestions as to how this important problem could be tangibly tackled in this modern-day technological society.

The Problem

The main objective of this study is to unravel the mystery that surrounds the 'problem' of discrimination, and to suggest an appropriate theory that could rightly explain the genesis of this 'negative human behaviour' in many cultures around the world. It is the firm hope of the present theorist that **a principle of non-discrimination law** should be possible to arrive at, and this should in turn help explain the existence of negative behaviour revealed by certain people in public as well as work places. This objective leads us further into the analysis of certain problems that exist in this physical world and have a bearing on man's relation with the natural environment. *If we define physics as the science of the contingent relations of nature, then the overall approach of this treatise inheres in the territory of physical science, where the "matter man" and his relationship with the physical environment are investigated.*

The central hypothesis is that discrimination arises out of the subjects' fear of contact with a foreign behaviour (organism) that it is not used to. In other words, to a large extent fear will be lessened if the subject, prior to coming into contact with this foreign behaviour (organism), is fed with more information about the latter that conditions the former. But my interest is not to explain the principle of non-discrimination by hypotheses, but to propose and prove them by reason, observation and if necessary by experiments; in order to which I shall premise the following definitions and axioms. For we must understand that a real definition may always serve as the premise, or part of the premise, of a logical inquiry concerning a subject matter. Thus from these definitions, together with our premises we can deduce the theorems which will follow later in this chapter.

Definitions

Definition 1

The concept of discrimination is defined as unfavourable treatment of the foreign organism (object) by the subject based on prejudice, especially regarding race, age, sex, ethnic or religious affiliation.

In connection with our examination, the definition will be said to entail the unfair treatment of foreign organism (migrants) by the subject due to his possessing one of the above-mentioned variables. Some relevant points need to be pondered over:

(a) In the case of discrimination there is the *awareness* in which the subject perceives the presence of a foreign organism (object) that the former is not used to, sees as strange, or is simply regarded as novel. There is a hypothesis that is associated with something that is novel, and this is called the *habituation hypothesis*. This hypothesis states that novel stimuli owe their collective properties to the fact they have not yet had a chance to lose effects that all stimuli originally possess. Obviously, all stimuli are novel at some time, and so all stimuli must at some time have the effects peculiar to novelty. But having once taken place, and especially having occurred repeatedly, they must lose these effects. How novel a particular pattern is will presumably be inversely related to (1) how often patterns that are similar enough to be relevant have been experienced before, (2) how recently they have been experienced, and (3) how similar they have been.
(b) There may be *motives* and *drives* that compel the subject to issue this negative exploratory behaviour leading to repel. Motives may not be easy to determine, as it may be associated with the subject's past accumulated experiences. Drives can be energised from the state of the mind or the apparent internal condition of the subject in that very particular moment.
(c) The *activity* of the discrimination that results in both the negative behaviour and utterances of the subject that indicates that the foreign organism (object) is simply not wanted in the vicinity, or does not "fit" in.
(d) *Identification* with certain ideologies, fables, and myths by the subject that reinforce the position of the subject that what he is engaged in doing is the right thing.

Definition 2

The concept "merit" comes originally from the Latin word meritum (which means "price" or "value"). The word "merits" denotes the quality of deserving well. When one speaks of merits, words such as "excellence" and "worth" come to mind. It could refer to a thing that entitles one to reward or gratitude.

In theology, the word is used to refer to good deeds as entitling to a future reward. It is also used for a person that deserves or is worthy of reward, punishment, consideration, etc. Nonetheless, in the academic circles, merit can encompass the individual's training he has received from an accredited university, with all its grades points as well as its certificates. It will include his work accomplishment in the different organisations, governments, and etc., and the education sector he has worked or served earlier.
Things to think about in terms of the application of merits:

(a) *There can be differences in merits.*
If at one point in time there is a candidate A who obtained the greater part of his merits concerning journalism through working in the banking sector as an analyst, it may be different from candidate B whose merits were gained in a book publishing sector as an editor.

(b) There can be differences in sources of merits.
The differences may arise when, say, one individual had acquired his educational merits through professional institutional training or studies and the other has secured the same kind of merits through ordinary apprentice training without any formal education as the former.

(c)Merits can be measured (the variables).
The number of years a candidate has spent in the classroom as a student, and the number years he been working in a particular branch of work, can be very important. If the ages of candidates can be considered relevant for certain jobs.

(d) Determinants of higher merits/magnitude.
Certainly a candidate with a university education (obtained Ph.D.) and several years of experience in the government sector as a politician has merits that are higher than a teacher training candidate with years of experience in elementary school. But here one can assert that it depends on the type of work being applied for, and whether they have also stated the type of qualification the job requires.

Merits denotes always the **time** (number of years) put in to study; therefore merits denotes not only time but also **quality** (e.g., published papers) of work accomplished, as well as the **energy** (strength and other resources such as money invested) put into one's accomplishments.

Definition 3

The word "quota" comes from Medieval Latin, which means "how great (a part)." In ordinary language it refers to "the share that an individual person or company is bound to contribute to or entitled to receive from a total."

It is also employed to refer to "a quantity of goods etc, which under official controls must be manufactured, exported, imported, etc." Or "the number of yearly immigrants allowed entering a country, students allowed to enrol for a course, etc."
Things to think about quota:
Determinants of the quota system: on what percentage should the quota system be based?
This determination can be made, and that depends on what the institution or particular organisation considers as fitting.

Definition 4

"Conditioning" may be defined as a response to a stimulus established by training. In the same vein, "conditioning reflex" is a reflex response to a non-natural stimulus established by training.

In a sense, conditioning can denote the manner in which a subject is gradually and systematically trained in order to aid him in learning about a stimulus that is foreign which in our study may refer to the foreign, organism (object). There are two kinds of conditionings: classical conditioning and operant conditioning. In respect to classical conditioning, some normally involuntary response is conditioned to a new stimulus. Regarding operant conditioning, what the person selects to do either brings some reward or takes away some upsetting situation. The results of the activity produce reinforcement, and as a consequent of reinforcement, behaviour alters.

Definition 5

"Impressed action" is the force exerted on the quota-merit principle by the state through its different agencies to keep the conditioning on its straight course.

The action consists solely in the impetus exerted by the super-ordinate organ (the state) on the agencies. There are various sources of the impressed force, such us Municipal Authorities, Regional Authorities and other Labour Market State organs. The necessary force or action is proportional to how heavy the strength of the resisting obstacle (opposition groups, underground groups). The required force is *zero* if the resisting obstacle stops acting as opposition.

Definition 6

"Resistances" refer to the various ideologies, myths, personal prejudices, and fables by which the subject uses to reinforce his position that what he is engaged in doing is the right thing.

Many of these variables have no basis except the fables and the myths, which I shall discuss below. They concern the false mythologies handed down about certain races through certain traditions since time immemorial. Certain academicians have coined other terms, but as for these we do not have to worry about them. Since they are the same old derogatory words said so many years ago such as "barbarians," "naked barbarians," and "the savages," and so forth.

Definition 7

"Impetus" means the manner by which the state overcomes the resisting obstacles presented to it by the underground groups and the opposition groups about its programme of changing the discriminating attitudes of the society in question.

Definition 8

"Inert state" implies the state in which the human organism is inactive in terms of employment. This is a situation whereby he finds his ability and resources as not being tapped in an adequate manner in the job market. This static condition, though full of energy, is perceived by the individual as very pathetic and needs to change.

Definition 9

"Kinetic state" is where the individual recognises the full utilisation of his potentials or abilities in the society. It is a moving state where his resources and energy are being demanded all over the market. It is a condition where his resources are perceived as marketable. It is a period where the individual can with certainty speak of actualisation of a potentiality.

Definition 10

Like electrons that carry electricity in solids, "minutrons" carry the charged mental ideas in the brain field, as well as transmit charges through wave mediums to another individual.

This comes from the words "man" and "neurons". *Minutrons* operates in the brain nerves, a fibre or bundle of fibres that transmits impulses of sensation or motion between the brain or spinal cord and other parts of the body. Nerve comes from the Latin word *nervus* and is related to the Greek word *neuron*. The letters "i" and "t" in *minutrons* represent international and telecommunication respectively. The complex ideas are usually charged ones. Pressure or indoctrination could easily accelerate them. These may make the individual feel that he is a "superhuman being" or an "omnipotent". Usually it concerns about superiority and the glorification of having some power over certain people or one's subjects. The individual erects a wall, and always sees himself on the other side of the wall or above high it. For instance, the experiences of the Ancient Babylonian King Nebuchadnezzar wanting to become like the most High and acting and thinking like the omnipotent, can be taken as a classic example. All is about superiority over others. It is possible that such infantile notions are associated with the individual not being able to have contact with reality.

Postulates

The following are the three postulates:

Postulate 1

As mentioned before in the prolegomena that we become more attentive to the recognition of the fact that there is a possibility to resolve the issue involving discrimination. Those negative behaviours, which constantly afflict the modern society as it has always done so many centuries ago, and had originated unwanted frictions among people and between different races.

Postulate 2

I postulate that an examination of the axioms or laws of non-discrimination and the propositions will provide us the necessary understanding, which can make us predict strongly about the assertion that it is possible for discrimination to be banished from the modern society.

Postulate 3

I postulate that it is important for mankind to observe meticulously and to judge for themselves the falsity or truth of these axioms. By doing so they will become capable and willing to readily admit the truth and certainty of the following axioms or the laws of non-discrimination.

Scholium

Thus far it has seemed best to clarify the senses in which less familiar words are to be taken in this treatise. Although discrimination, action, quota, merits, conditioning, impressed, action, force, kinetic, inert, impetus, and resistances are familiar to everyone, it must be asserted that these terminologies are popularly conceived solely with reference to other subjects' areas and not like the manner I am currently employing them. As this can be the source of certain misconceptions, it is essential that I discuss further to eliminate any misconceptions so as to make them useful in our present investigation.

While discrimination, action, quota, merits, and conditioning are employed commonly in the behavioural sciences, impressed, action, force, impetus, kinetic, inert, and resistance are terminologies found commonly applied in physics and astronomy. And in this treatise these terminologies have been used almost similar to the meanings that they are frequently employed in astronomy and physics. Thus in the field of astronomy, resistance is used to refer to the manner in which air or gravity impedes the progress of projectile when it is thrust into space in motion by a human subject. The projectile, by gravity, is drawn from a rectilinear course and continually deflected toward the earth, and this is so to a greater or lesser degree in proportion to its gravity and its velocity of motion. Impetus means, insofar as the same body, yielding only with difficulty to force of a resisting obstacle, endeavours to change the state of that obstacle. Resistance is commonly attributed to resting bodies and impetus to moving bodies. But in our work, resistance is employed as performing a similar function whereby the myths, fables, and personal prejudice impede actions of the human subject to discourage proper appropriate acceptance to occur between him and the foreign organism (object). Resistance prevents the mental faculty to issue appropriate behaviour toward the foreign organism. Impetus, on the other hand, means the manner in which the state overcomes the obstacles presented to it by the underground groups and the opposition groups about its programme of changing the discriminating attitudes of the society in question.

Astronomers employ action or force to mean the manner in which the heavenly bodies are made to orbit the stars through the impressed action that equips them with motion. The various sources of impressed force are percussion, pressure, or centripetal force. One force of this kind is gravity, by which bodies tend toward the centre of the earth. Another is magnetic force, by which iron seeks a lodestone, and yet another is that force, whatever it may be, by which planets are continually drawn back from rectilinear motions and compelled to revolve in curved lines. However, in our treatise, impressed actions are denoted by the state or a super-ordinate power that accelerates the quota-merit principles that are used to determine the priority of acquiring jobs for the foreign organisms (*subjectuses*); this puts discrimination in check. Thus force and action are used interchangeably to mean the government action or role in matters regarding discrimination. Kinetic and inert comes from the kinetic theory of matter that explains the physical properties of matter in terms of the motions of its constituent particles. Kinetic thus means energy which a body possesses by virtue of being in motion.

Inert is a property of matter by which it continues in its existing state of rest or uniform motion in a straight line, unless that state is changed by an external force. In our treatise, while the latter represents the inactive state of the organism in the job market and the society as whole, the former indicates the state where the organism sees enormous potential use of his ability and personal resources. In fact, in the kinetic state one can speak of *actualisation of a potentiality* of the individual.

In the behavioural sciences, such as economics, sociology, education, and demography, otherwise known as population studies, discrimination, quota, and merits, appear frequently. The uses of these terminologies in these disciplines correspond to the manner I employ in my treatise. It appears, therefore, that one should not be confused with them as they are commonly employed in diverse areas. Discrimination denotes the general approach of sorting out people according to colour, religion, race, and sex, and it becomes the canon where people are given jobs or refused in certain societies around the globe. Even the language one speaks can make an individual become discriminated at certain public places or work places.

Conditioning is a psychological terminology used frequently in the study of behaviour. Early employment of this terminology was in the research among animals such as dogs, but later it was transferred to human beings. Ivan Pavlov, a Russian Psychologist, pioneered this. Later B. F. Skinner, a psychologist in the USA used it in his investigations. Conditioning is considered the simplest form of learning.

In classical conditioning, some normally involuntary response, such as the blink of an eye, is conditioned to a new stimulus. For example, scientists investigated the eyeblink response in men and women ranging in age from 18 to 85 by having each individual sit in a cosy chair, while speakers mounted about a foot from each ear sounded a tone just before a puff of air were blown into the eyes. When adults blinked at the sound of the tone, even though no blast of air followed, they were conditioned. Young adults conditioned rapidly, as did those in their forties, but after the age of 50 the speed of conditioning slowed dramatically and the proportion of adults who eventually developed the conditioned response dropped. In operant conditioning, what the person chooses to do either conveys some reward or removes some unpleasant situation. The consequences of the action of the individual produce reinforcement that, and as a result of reinforcement, behaviour changes. The effectiveness of operant conditioning in modifying human behaviour can be seen in its utilisation to the case of an 82-year-old patient with heart illness who regularly stayed in bed all morning, scarcely exercised, did not take his medication, and refused to drink his orange juice. Each time he walked around the block without being reminded, drank a small glass of orange juice, or took his medication, he earned tokens which could be exchanged for such privileges as a weekend dinner at a restaurant of his choice. Within a few weeks, his walks had increased from once every seven days to three times a day, he was drinking three glasses of orange juice a day instead of one—or none—and he was taking each of his three medications regularly. During the same period his angina pain disappeared.[2] Both the classical and operant conditionings are employed in the formulation of the theory of non-discrimination. Conditioning is used in the Psychiatry Hospitals, even psychotic older patients and patients with organic brain syndromes have been successfully conditioned.

In what follows, a fuller explanation will be given of how these terminologies function to provide us with a solution to the problem of discrimination.

---

[2] See Perlmutter & Hall, (1992:216-217).

Chapter 2: The Principle Of Non-Discrimination Laws, Or Axioms And Other Propositions

Proposition 1

There is no such thing as discrimination toward a foreign organism (object) in terms of the seeking for employment (sociability) in any society, but the fear of the foreign organism (object) causes the negative repels due to the lack of sufficient knowledge by the subject on the side of the foreign organism.

If the subject is aware of the foreign organism and has established through a considerable amount of time a basic relationship, and moreover has knowledge of the latter there will be acceptance (Def. 4 and Law 1). Fear, in this case, will *dissipate* and there will be more time to learn about each other (Def. 1). The hidden psychological fear that the subject has does not mean that all things are bad from the viewpoint of the proponent of this proposition. It will be established later in our analysis that even **the strategy of maltreating the foreign organism through the use of mobbing or temporary rejection depicts a sign of extreme hidden fear of any organism that engages in it** (Def. 8). It shows lack of self-security on the side of the subject (Post. 1).

Proposition 2

If the fear of this foreign organism (object) the subject is not used to is removed temporarily or permanently, the subject will be more receptive to the foreign organism.

It is the removal of the fear that is presenting a block to a better relationship that will give course to celebrate for successive interactions (Def. 4 and 5).

Law 1

If you condition the subject who, as a result of the fear of the foreign organism (object) is not capable of relating well to the latter, it will not produce discrimination, which is equivalent to the inducement of conditioning produced by the combined application of the quota system and the merit principle.

It is the fear and contact with the unknown, the strange, or the novel that trigger the subject to commence treating the foreign organism the way the former does. The former will not like to have much contact with the latter, and in terms of employment he will not offer even if the latter possesses the necessary documents and qualifications. Whenever the condition methods are applied, it is supposed to provide the subject appropriate way of getting to know the foreign organism better in preparation for true acceptance. Law 1 states that to utilise the condition methods individuals come to better *adapt* to the situation of living with other persons, which is learned in a gradual and systematic manner. Meanwhile, where the combined use of the merits-quota measure (impressed action) have enabled people through adaptation to be easily accepted in work places, conditioning will be equally able to perform the same function without needing to resort to the quota-merit principle.

Law 2

The quota-merit principle is a form of an inducement of conditioning and it is related (proportional) to non-discrimination through the impressed action/force.

To use the quota-and-merit principle to secure employment for the foreign organism to extent becomes a matter of conditioning process. Though the subjects may be temporary opposed to the idea, eventually they will learn or be trained to get used to it. Its use continuously will ensure that both the foreign organism and the subject learn to live with each other. The use of this measure is possible so long as the force or the action is impressed by a superordinate power from external of the organism and the subject; and is tantamount to saying/ensuring that discrimination is banished permanently.

Corollary 1

With the employment of the conditioning method on the subject that fears the foreign organism, the amount of the non-discrimination behaviour that is produced in the subject has an equal effect as the inducement of conditioning produced by the combined application of the quota system and the merit principle.

Common sense teaches us that if you through a gradual and systematic manner introduce the foreign organism to the subject, the latter will not fear the former; in fact, eventually the acceptance will be deep and strong (Law 1 and Def. 4). There may be emotional or friendly attachment which will result; either the subject may form a closer relationship or bond leading to a better orientation, or through the exploratory phase come to possess *agape* friendliness. The quota-merit principle has similar effect, in that though during the commencement the action of putting a foreign organism there (impressed action, Def. 5 and 7) will be abhorred, gradually the subject will open doors for acceptance to the foreign organism as learning and illumination gain weight over/overcome the resistances (myths, false ideologies, fables and personal prejudices, Def. 6).

Corollary 2

The use of the condition method in a prolonged period of time/perspective will make null and void the combined application of the quota system and the merit principle as the inducement of conditioning.

Time is needed for healing to take place; time is also indispensable for teaching to gain root in the individual as well as time enables people to acquire or possess patience (Def. 1). On the relationship between the strength of a conditioned stimulus and the magnitude of a conditioned principle, with time, it is possible to predict that the mechanisms involved in the gradual and systematic learning will make the quota-merit (force action) null and void (Def.

5). In other words, people's mental, habits, emotions and taste change with time. Learning and training have positive results as the organism after graduation is not considered to be the same person as before (when he/she commences her studies).

Corollary 3

When the organism is conditioned it removes any resistances, these being some of the internal impeding forces that are enshrined in the motive, and they include false ideologies, myths, personal prejudices, that which cripple the mental faculty of the organism (i.e., human subject).

Let us say that subject A possesses a fearful disposition toward foreign organism B (the object). Usually there are some resistances that make it very difficult to approach or have social interactions with the latter (Def. 6), and these are often present with the former in the form of false ideologies or doctrines, myths, and personal prejudices imbibed during childhood. According to Law 1, the removal of these resistances is imperative for the subject to come to a "state of equilibrium" that implies complete acceptance of the foreign organism in his vicinity (Def. 9 and 7).

Corollary 4

Conditioning the organism can be regarded as introducing a certain form of (natural) mental gravity that, when once set in motion, does not need to change with respect to time.

Any adult human organism possesses in his mental faculty the idea of rights and wrongs that guide him in the manner he relates to other fellow human beings. If the organism is helped to understand properly what the purpose of this conditioning is about, its influences on the mind with respect to time will not diminish. In other words, mental influences once impressed on the human organism are, in it self, indelible, and cannot be changed or rubbed away easily (Def. 5).

In what follows I shall discuss the origin of fear, and then come to touch upon how the theory of conditioning that has been tested on different organisms could be utilised to elucidate the principle of non-discrimination law.

Awareness and fear

In real life situations the strange, the unfamiliar thing or object, the novel, is apt to repel and to provoke the emotional disturbances that social psychologists as well as behavioural scientists consider as indexes of fear. The same phenomena in other circumstances may elicit indexes of pleasure, attract animals toward them, and may be eagerly sought out. "The properties of objects or situations that we have seen to be most potent in prompting exploration resemble those that we might expect to provoke 'fear of the unknown.' Thus, "we might thus expect novel stimulus patterns to attract and repel in turn or to arouse some degree of conflict or vacillation between approach and withdrawal."[3]

To take a common illustration that was given by Berlyne in his work so many years ago, let us say that an unfamiliar object, such as a human hand, is suddenly pushed into a rat's cage, or a novel toy is thrown in front of a baby. The contest between the two opposite tendencies, according to Berlyne, is likely to be distinctly perceptible. From quite a distance the individual subject will examine the intruding object, "apparently alert and delicately poised, perhaps even oscillating, between advance and withdrawal." In case the object shifts in a different direction or sideways, or probably makes a sudden alteration, the subject will be alarmed and this will be followed by a quick retreat. But if the subject remains in a static position while pretending to appear harmless, the subject will sooner or later commence to edge gradually forward till it can be closely inspected.

Even among birds Hinde[4] and Marler[5] have observed a similar kind of behaviour. Chaffinches are said react to the sight of a predator by issuing "mobbing" tactics, a characteristic response pattern that includes advancing forward to within a few feet of the predator, making short flights alternately toward and away from it, and uttering peculiar "chink" sounds. Mobbing tactics in this case is somehow a mysterious behaviour, though it is thought to serve as a threat to the predator. But mobbing has been found to be not always an effective approach to frightening a predator, as it could also make chaffinch more vulnerable to attack than it would have been otherwise. Other chaffinches may be warned or lured away from chaffinches' young by the issuance of mobbing. According to suggestions put forward by Hinde and Marler, mobbing is at least partly a form of exploratory behaviour, "vacillatory movements reflecting a conflict between approach and escape." Thus, in Berlyne's words, a tendency to observe sources of danger from a distance makes good biological sense.

One prominent component of mobbing is visual fixation of the predator with the medial fovea of each eye in turn, and mobbing exhibits the same short-term and long-term decrements with prolonged exposure to the eliciting stimuli that have been found in the exploratory behaviour of the rat. In Marler's reports it is noted that chaffinches are in the habit of responding to any strange object with behaviour that closely resembles mobbing, and then provides a "subjective impression of curiosity or inquisitiveness." When this has been done, they certainly appear to "avoid (the object) or its location for a time."

Flight can be an exploratory strategy

The research of Lorenz[6] provides us a vivid description of the context between flight and exploratory manner in raven:

A young raven, confronted with a new object, which may be a camera, an old bottle, a stuffed polecat, or anything else, first reacts with escape responses. He will fly up to an elevated perch and, from this point of vantage, stare at the object literally for hours. After this, he will begin to approach the object very gradually, maintaining all the while a maximum of caution and the excessive attitude of tense fear. He will cover the last distance from the object hopping sideways with half-raised wings, in the utmost readiness to flee. At last, he will deliver a single fearful blow with his powerful beak at the object and forthwith fly back to his safe perch. If nothing happens he will repeat the same procedure in much quicker sequence, with more confidence. If the object is an animal that flees, the raven loses all fear in the fraction of a second and will start in pursuit instantly.

---

[3] Cf. the work Berlyne (1960).
[4] Hinde (1954). Factors governing the changes in strength of a partially inborn response as shown by the mobbing behaviour of the chaffinch *Fringilla Coelebs*. I and II Proc. Roy. Soc., 142, 306-331, 331-358.
[5] Marler (1956).
[6] Cf. Berlyne (1960).

If it is animal that charges, he will either try to get behind it or, if the charge is sufficiently unimpressive, lose interest in a very short time. With an inanimate object, the raven will proceed to apply a number of further instinctive movements. He will grab it with one foot, peck at it, try to tear off pieces, insert his bill into any existing cleft and then pry apart his mandibles with considerable force. Finally, if the object is not too big, the raven will carry it away, push it into a convenient hole, and cover it with some inconspicuous material.

A careful observation of Lorenz's extract reveals how the raven's exploration of strange objects is combined with fragments of behaviour connected with eating and fighting. Even experimental studies performed later after Lorenz's work with the employment of monkeys and cats has revealed that stimulation of most cortical points resulted in both implicit signs of the orientation reaction and overt orienting responses. But when certain points were stimulated, these responses were combined with alimentary or defensive responses.

Three Russian scientists, Dolin, Zborovskaia and Zamakhover,[7] concerned themselves with the context between *caution* and boldness in the reactions of monkeys and apes to strange stimuli. In their experiments they provided various objects; that is, geometrical figures, tin boxes, toys, and smaller animals, into the living cages of their subjects and observed from a close distance on a one-way screen. The first reaction was invariably one that these researchers call the "inhibitory reaction." The subject kept his distance from the stimulus object while remaining frozen in one posture, staring and often gaping. What might be referred to as the "first phase" gave way to a "second phase" that involved active exploration. Here the subject approached the object, scanned it from all sides, sniffed at it, touched it and handled it. The experimenters stated that there were often traces of alimentary, aggressive, or sexual behaviour. In short, the exploratory phase might last for over an hour, but if the strange object continued to be present, periods of active exploration would alternate with periods of passive starring in a trance-like state. With certain individual subjects the inhibitory reaction was more likely to be prominent especially those with living stimulus objects and on the first day that the object was brought into contact.

Device for relieving fear

Grave fear can sometimes be surmounted during the exploratory phase of the novel when the subject has access to a device for relieving the fear. This was demonstrated in the research carried out by Harlow and Zimmermann[8] with infant monkeys. They note that an infant monkey brought up by a different mother who is not her natural mother and in the presence of a cloth-covered model will come to regard the model as a mother substitute. When the animal is put in a room that is filled with strange objects but without the substitute model, it is likely to "freeze in a couched position" or "run rapidly from object to object screaming and crying." But if the model is brought into the room, the animal will cling to it. Continuous tries with the mother substitute lessen or eliminate the fear entertained by the animal, and this will bring the exploratory tendencies to its consciousness. The monkey will utilise the mother substitute as a "base of operations" and "will explore and manipulate a stimulus," and then make back to the mother before adventuring" again into the novel fresh milieu.

The bulk of study that has been conducted on living organisms and the fact that whether exploratory or fearful reactions will come to the fore when they come into contact with a strange pattern of stimulation demands further exploration. But for us, these studies briefly analysed provide us enormous information regarding how living organisms react to unusual pattern of stimulation that is near to the manner of which living beings as a whole

---

[7] Dolin, A. O., Zborovskaia, I. I. and Zamakhover, S. K. (1958). On the characteristics of the role of the orienting-investigatory reflex in conditioned-reflex activity.
[8] Harlow, H. F. and Zimmermann, R. R. (1958).

encounter something that is *unknown, strange,* or *not used to.* In the case of the animal subject, it seems the matter is with how novel or complex the stimulation is. Extreme novelty or complexity, as studies have unfolded, tends to induce avoidance, while moderate novelty or complexity all the same induces a certain amount of approach that is still careful. However, one factor that apparently makes a distinguishable difference is whether a subject is suddenly and with force put into the midst of a totally unfamiliar milieu "or whether its environment contains both relatively novel and relatively familiar elements."

Referring to the last-mentioned case, the animal subject generally has a tendency to observe a distance from the strange things at first, with perhaps a few scattered exploratory forays, and then become more and more inclined to expose itself to novelty as time goes on. *"But when an animal is attacked by strange stimuli from all sides, it does not have the same choice."*[9] The intensive exploratory behaviour will be initiated first, though it will suddenly decline rapidly. Then the exploratory activity will be replaced by behaviour indicative of fear, as the work of Bindra and Spinner, through their meticulous observation, has documented. Without observing whether it gives way to a tense or a relaxed state, other researchers have been satisfied only to pen down the decrease that occurs in exploratory behaviour. Here one notices that "the fear is, of course, occurring at a stage when the animal would not yet have been ready for much spontaneous approach to the novel stimulus objects if it had had the alternative of remaining in familiar surroundings."

Perceptual curiosity, learning and conditioning

Perceptual curiosity may be removed during the involvement in the conditioning or the learning that the subject acquires. Moreover, one cannot doubt the role motivation plays in the exploratory behaviour of the subject in this "sequence" of learning in general. Woodworth[10] demonstrates this in experiment. The experimenter recognises a phenomenon that he calls "sequence learning." This occurs when a stimulus, S1, is frequently succeeded by another stimulus, S2, and the subject learns to execute some response in preparation for S2 during the interval between S1 and S2. He does not claim that all learning conforms to this paradigm, but it seems to fit classical conditioning advocated by Pavlov, instrumental conditioning (operant) propounded by Skinner, and paired-associate rote learning, among other instances.

The process commences with S1 which, according to Woodworth, evokes in the subject's mind something that sounds like a question: "What's that?" or "What's next?" or "What does S1 portend?" This arouses a "questioning set," "a readiness for some unknown S2," an "indefinite expectancy of something more to follow." As soon as S2 appears, it provides the answer to the imaginary question and changes the indefinite into something definite, and this stage of the process gives the reinforcing factor that establishes the learning.

The importance of Woodworth's study is that it makes it easy to identify the state that S1 induces, as there is an increase in arousal and in fact this prepares the subject for the removal of perceptual curiosity by the receipt of S2.

Vital notes on human behaviour and conditioning

---

[9] Italic is mine.
[10] Woodworth, R. S. (1958).

Students of classical conditioning, as Anokhin[11] informs us, have noted that the original reaction to the conditioned stimulus usually consists of orienting behaviour that is replaced, as reinforced trials follow one another, by conditioned orientation in the direction of the unconditioned stimulus and then by the specific conditioned response. When an animal, for example, has to learn to press the bar in a skinner box in an instrumental conditioning situation, the reaction first takes the form of sniffing about and when a few rewarded bar pressings have happened, especially in the neighbourhood of the bar. This exploratory behaviour vanishes as the instrumental response becomes firmly rooted. And in more complex situations, as soon as the animal obtains, and has to respond correctly to, successive stimuli, the various links in these successions first elicit separate orienting responses, and then orienting responses to all except the last link gradually drop out.

Studies that concern the function of verbal responses in those forms of behaviour that can be regarded as "conscious," "voluntary," or "rational," have been carried out. It is important that the findings in these areas be made known though they are rather old, like most our data on conditioning. The work of Lisina[12] reveals how the researcher managed to train human subjects to control their normally involuntary and unconscious vascular responses with the help of orienting responses and selective attention directed to appropriate cues. The subjects learned to dilate the blood vessels of the hand in response to a shock, a reaction running counter to the natural defensive vasoconstriction response, and to produce vasoconstriction and vasodilatation at will in response to instructions from the experimenter. The former feat was done by allowing subjects observe the recording pen of the plethysmograph, whose fluctuations, the subjects were informed, represented some aspect of the working of their bodies, and then switching off the shock as soon as vasodilatation happened. The latter was done by a course of training in which subjects had vascular reactions explained to them and were taught to distinguish the sensations coming from their own vasoconstriction and vasodilatation.

The analysis will not fail to mention numerous researches that have been conducted among the young and older human subjects by Solomon and his associates (1989), Woodruff-Pak (1990), Rinke and associates (1978), Balota, Duchek, and Paulin (1989), and McDowd and Brien (1990).[13] When strange people we do not know say "hello" to us on the street or greet us, we are taken aback and often ask ourselves, "where does this person know me from?" But human beings, unlike animals have developed certain ways of responding to such unwanted advancement, and we may have been naturally conditioned not to show these surprises or fear in public. One simple way is not to allow the organism a chance to come near or even have the opportunity to socialise with one, for we are still afraid by the fact that by opening up we may be tempted to give up our false ideologies, fables, personal prejudices, or myths we have imbibed since childhood. It is not the encountering of these resistances alone that prevent us from coming closer, but rather the psychological pain it involves in dealing with these resistances when at last we discover that these are all untrue about the foreign organism. "They are like us," "they reason like us," but it is true we have lived in a different environment and our problems and difficulties we had encountered so many thousands of years ago have not been the same. This is where the difference lies, but otherwise we have many things in common though we are different in many ways due to the problems we have encountered. Let me leave this for now and go on directly to the scholium of the theory to offer some discussion on the subject.

---

[11] Anokhin, P. K. (1958).
[12] Lisina, M. I. (1958).
[13] Permutter, M. & Hall, H. (1990).

Scholium

Thus, according to the principle of non-discrimination law which I have meticulously developed above through deliberative discussions and appropriate definitions, systematic or unsystematic conditioning of the fear of the subject will not produce discrimination toward the foreign organism (object). This is equivalent to the conditioning induced by the combined application of the quota-merit principles (i.e., the merits received a quota or "the merit was quotas"). Let us say that these letters below represent:

Box 1: Formulae of the principle of non-discrimination law.

C represents (conditioning)
F represents (fear)
$đ$ represents (no discrimination)
Q represents (quota system)
M represents (merits)
α represents (total action or work done by the state in changing the discrimination attitudes)

Then the formulas or **the equations** for the principle of non-discrimination law which I am proposing are:

Law 1 $\Leftrightarrow$ *Con. f* $\leq [đ = (qm)]$

Law 2 $\Leftrightarrow đ = qm\alpha$

Merits (**m**): This always has to do with *time* (number of years of study in addition to number of years worked) put in to study; therefore merits denote time and *quality* (e.g., published papers) of work accomplished as well as *energy* (strength and other resources e.g., money invested) put into accomplishments. Therefore m = (t×q×e).

Quota (**q**): This is decided in *proportion* to the total number of people in a community. Proportion to the total number of workers needed for a particular industry. If you increase the total number of workers, you must increase the number of *subjectuses*. For example, if Q = (20/100 × 56) then 20 is the percentage of the *subjectuses* divided by 100 multiplied by 56, that is the total number of people to be employed.

No discrimination (**đ**) $\leq$ fear diminishment (no discrimination is "equal to or in the order of" the *dissipation* of fear).

Action (α): The necessary force/action is proportional to how heavy/the strength of the resisting obstacle (opposition groups, underground groups). The required force is *zero* if the resisting obstacle stops acting as opposition. On a scale, the State = **3**, the Municipal Authority = **2** and any other Organ = **1**. Where 3 represents the highest action or force applied.

The principle I have set out follows behavioural scholars in the field of psychology and other behavioural sciences who have carried out considerable studies to unravel the mysteries about human behaviour or personality. Usually the theories they come up with are accepted to be authentic and reliable. Conditioning theory already gained popularity in the 1950's, and since then a greater amount of research regarding animals, birds, and human behaviour or personality have been conducted in this area. Many pioneering scholars have employed this theory to solve human problems in different ways. For example, Ivan Pavlov, the Russian Psychologist, and B. F. Skinner, the American Psychologist, spent the greater part of their research time on the conditioning methods on both animals and human organisms.[14] Skinner, for instance, at his old age in 1983, conducted research with himself and told how he used these methods to outwit his ageing memory. Solomon[15] and his associates, Woodruff-Pak,[16] Rinke[17] and associates, Balota, Duchek, and Paulin,[18] and McDowd and Brien,[19] are just a few of the countless number of scholars who have showed that conditioning methods are authentic and could be helpful in modifying human behaviour.

Furthermore, what I have stated as laws regarding human behaviour and their corollaries may be applicable to the universe of discourse regarding all human behaviour. It is the firm hope of the proponent of these laws that, given the chance, some, if not all, of the majority of people will behave fearful toward anything they are not used to or consider as novel and strange. This is the basis of the principle of non-discrimination law. The arousal of fear in this case has not only been studied with human subjects, but also studies utilising animals and birds have revealed that contact with the unknown, novel or the strange causes some concern to *all living organisms in this universe*. But when the organism is continuously bombarded by strange stimuli from all sides (in the case of our work, influx of immigrants), it does not have the same choice. Then a considerable amount of time is required for the subject to engage in exploratory behavioural study of the foreign organism or object it encounters. As the subject is taught through conditioning to have systematic knowledge and consequently direct acquaintance with the novel, strange, or unknown, fear dissipates, and this in turn leads to warm acceptance, which offers tranquillity and equilibrium mind to the subject and also the foreign organism. At this time a better relationship can develop to a superordinate level, and a kind of acceptance which may cause the subject to relinquish some wrong impressions (i.e., wrong ideologies, myths, and group pressure) that he might have entertained at first about the foreign organism will result. A kind of *elementary* trust, which has never been given before to the foreign organism but necessary for any developing relationship, will be given to the foreign organism from the subject.

For the purpose of practical solutions to the fear of the subject of the foreign organism, I say that the propositions in this investigation are indeed correct and demand a further study using empirical materials collected in the field. In accordance with my meticulous observation in the field and the tremendous experiences gained as a result of my superb human interactions, I propose the following solutions. These are temporary, to be considered as conditioning methods or approach that will help dissipate the fears of subjects beset with the problem of discrimination in work places. Firstly, the *soft conditioning incentive* (to use the Latin word for stimulus), and then secondly, the *hard conditioning incentive*.

---

[14] Permutter & Hall (1985).
[15] Solomon and his associates (1989).
[16] Woodruff-Pak (1990).
[17] Rinke and associates (1978).
[18] Balota, Duchek, and Paulin (1989)
[19] McDowd and Brien (1990).

The soft conditioning incentive should consist of:

(1) If it is necessary and appropriate, organisations as well as companies should be directed from the inception by the municipalities as to the proportion of foreigners against citizens (native) to be employed by them in their work.
(2) If it is necessary and appropriate, that the nation's television channels should regularly show how immigrants who have successfully obtained jobs carry out their work routines successfully.
(3) If it is necessary and appropriate, important newspapers in the country in question should devote one or two pages where they only publish the organisations and companies who have employed the proportion of the immigrants required by the municipalities.
(4) If it is necessary and appropriate, programs that compare the experiences of the country in question and other countries should be shown regularly. Both the negative and positive sides must be shown.

The hard conditioning incentive should consist of:

(1) If it is necessary and appropriate, an action/force that is more affirmative should be utilised: Particularly, parliament should vote a decision that will stipulate employment based on the proportion of the immigrants in a particular city, area, or region.

An action or a force is when you through legislation or peaceful means, compel citizen A or B to do something that you believe is beneficial for him/her and the society. Human nature sometimes need a kind of action or force to make it accomplish a task; and I say, if sometimes a force is what is required to modify human behaviour that is detrimental to society, then let it be used.

Impetus Principle

I demonstrate the impetus principle presented above briefly as follows. Suppose that at a given point in time the state **A** has applied the principle of non-discrimination law so that the foreign organism **B** will be able to have a position somewhere in the society in question. A third party, **D,** that is the underground groups may attempt by different methods so that this principle may not be successful or be able to carry through. This stronger pressure will not be allowed to prevail, since by the principle of impetus the state **A** will do its utmost best to overcome this resisting obstacle or endeavours to change the condition of that obstacle. This follows from Law 2 and Def 7.

        Corollary 1. Hence the impetus principle, when applied in the non-discrimination endeavours, manage to put the pressure of the underground groups in a balance that follows the equality of action and reaction law.

        Corollary 2. If the pressure comes from the superordinate force, that is the state, it will deter the underground groups, and this will signify some seriousness of the government to eradicate the inappropriate behaviour regarding discrimination.

Induction Theorem

If the state fails to utilise its impressed action and the employment of the quota principle, I say that the proportionality of the merit principle alone to fight discrimination will not function.

The derivation of this theorem (Law 2, Def 5) is vital, as it describes the situation that is prevailing in the job market at present. Many people have already obtained the necessary merits and all that which is basically need to aid them to acquire positions in the working sector. But this is not functioning as expected, because the forces in the market work as opposing groups to derail the plans to fulfil the integration programme. So simply put, the merit principle alone is insufficient to put the *subjectuses* to different jobs in the market.

Therefore, to apply the merit principle alone to acquire positions, which would have been appropriate under normal circumstances, is nearly equal to zero. The implication is that only the implementation of impressed action together with the quota principle will enable people to secure some positions in the labour market. It is only the equality equation between non-discrimination and merit-quota principle induced by the state that will alter the situation in the market.

Let $đ$ be the non-discrimination, and m be the merit principle, and α and q represent the impressed action and quota respectively. Then the formula for our theorem can be written thus where α remains as constant:

$$đ = m q \cdot \alpha$$

$$m = \frac{đ}{Q} \alpha = 0$$

Omega Theorem of the Principle of Non-discrimination

The theorem states that: *Increasing the non-discrimination momenta concerning employment is proportional to the intensity of the conditional strategies, and its success is dependent on the impressed action from without and the removal of the resistances.*

This follows from Law 1, Law 2, Def. 6 and Corollary 3.

Here we introduce a new concept concerning discrimination—*the non-discrimination momenta.* This is the random fixing of some *subjectuses* in the different parts of the working sector in the society pioneered by the state agencies. *Conditional strategies* refer to the methods that are utilised for the purpose of conditioning. The demonstration of this theorem is important in any society that intends to rid itself from the destructive behaviour of discrimination. It states that while propagating the strong idea of solidarity to ensure that citizens co-operate on the issue involving the dismantling of discrimination, the state should intensify its grips on the strategies that condition people to assist them from practising this negative behaviour. In fact, the increase of the former should lead to the increase of the latter; and all these should be the sole work of the impressed forces that include the state and its agencies. While these are going on the removal of the myths and fables that act as resistances, they should be tackled or they should endeavour to throw them away. The Omega theorem naturally suggests itself of concentrating one's attention also on individual cases on the dismantling of discrimination. An action in this domain is imperative in the field of discrimination that fear is seen to be associated with, and is also seen as the major contributor.

Thus, if εC be the intensity of conditional strategies, α the action of the impressed force, r the resistances, and $Ð_{\ni}$ the non-discrimination momenta, we can put forth the formula of this theorem as follows:

$$Ð_{\ni} = \varepsilon C\alpha - r$$

Where the non-discrimination momenta is described as related to the intensity of the conditional strategies by the impressed action without the resistances acting as barriers or hindrances.

Corollary 1. If the resistances are removed and the non-discrimination is employed, the work of the impressed forces will be made easier and the conditional strategies would be capable of functioning appropriately.

Corollary 2. Hence the banishment of discrimination is indirectly associated with the removal of the resistances that act as barriers or hindrances for genuine acceptance.

Lucifer Theorem

With the aid of the above theorem, the forces of *the underground pressure* can be readily felt. *Let the impressed forces be active in its dynamism in implementing the condition strategies as well as in the removal of the resistances, it will awaken the opposing reactions of the underground groups.*

This can be easily stated as **equality of action and reaction**. It follows Def 1, Corollary 3, Def 7, Def 10, and the impetus principle. Nothing is so obvious as the work that the underground groups constantly do to maintain the counter equilibrium situation as it does at present. An action by the state to change the situation will always meet a reaction from them in different quarters both in the higher level as well as the lower level. At the higher level are those influential people who sponsor the underground organisations, and at the lower level are those citizens that suffer deprivation of some kind. The latter actually do the work of causing distractions and acting as opposing groups. This equality of action and reaction is vice versa, as the state usually strikes back whenever it sees these groups causing extreme trouble or intimidation to the innocent people.

Corollary 1. Therefore, an impressed action is always needed to oppose the equality of action and reaction principle that is inherent in the field of discrimination.

Corollary 2. If the state strikes back constantly, then this will minimise the destructive forces working against the action of utilising the non-discrimination momenta.

Sola Fide Theorem

If the impressed action originates from the ultimate superordinate agent, I say that the quota-merit principle will condition the society's citizens for a complete mature acceptance of non-discrimination law, as individuals will become secured and this will ensure a distinct awareness of its implementation that is comparable to a healing process.

This theorem follows from Law 1, Law 2 and Corollary 4. For if the highest power in the land or a country is the one pushing this conditioning that permits the quota-merit principle to be implemented, it will aid this execution to enable some latent conflicts to surface. As the highest authority in the society in question is involved, these conflicts will appropriately be managed or dealt with in a mature manner nation-wide. The consequence is the ensuing of solidarity that is seen as positive in the governing of any nation. The nation in question will witness the positive result of progress that is concomitant with solidarity, advancement, and economic development.

There will be renewal of social relationships, and this will make people to adhere to each other in a meaningful way. The result will be the renewal of confidence in the ruling government of the state and its agencies on this particular issue being discussed. In short, there will be **faith** in the authorities that can be attributable to the social integration plan of the government being capable in dealing with the national problem in an efficient and a proper manner.

Corollary 1. If the highest agent in the society champions the implementation of the quota-merit principle, I say that it will provide a healing experience to the nation's citizens.

Corollary 2. Hence the quota-merit principle can heal the wounds of a nation.

Scholium

The four theorems the Induction theorem, the Omega theorem, the Lucifer theorem and the Sola Fide theorem, all informs us concerning the necessity of the state as the highest organism to lead in this essential work of the diminishing of discrimination in the society in question. We also deduce from the axioms, which indicates that the implementation of this important work has some advantages that will accrue to the power and confidence in the state. This will, therefore, be seen as a boost to the state power in carrying out other similar functions or assignments in the society.

But my purpose here is not only to present these propositions and research in connection with the laws on non-discrimination, but also to build a complete theory concerning some other principles that can help in the dismantling or disentangling of discrimination. By these examples I wish to build an entire theory that will help discrimination to be banished from the modern society. To make a deeper discussion of these two conditionings incentives proposed above, first I shall speak of the former, which is the soft conditioning incentive.

Chapter 3: The Soft Conditioning Incentive

The soft conditioning incentive would work well in areas such as the Post Office, public sector jobs, government companies, universities, secondary schools and elementary schools, agricultural sectors, and the manufacturing sectors. If it is necessary, the soft conditioning incentive should be directed at government works or the public sector in the first instance. Later it can then be shifted to other private sectors in the society in question.

Special conditions regarding these areas are that positions become the property of some employers and they find it very hard to let a vacant position go to an immigrant. Even the ordinary native worker/employee that has just been employed becomes a boss in his own right, and sees the threat of the immigrant a big worry. This condition is registered in the unconscious mind and therefore people usually feel secured to employ their close relatives first, followed by closest friends, and then people they are acquainted with before they will take ordinary citizens they do not know. The case of the immigrant becomes a matter of difficulty, as here also they look for someone they either know through friends or celebrities before they come to the ordinary immigrants. Usually they look for loop-holes that could disqualify you for the jobs, therefore any little mistakes one makes during an interview could lead to one being rejected. Like, for example, using an e-mail to write an application instead of an ordinary letter, though they have indicated that it was acceptable, this can be a disadvantage to an applicant. Or make a little sign that indicates that one's language is not good could also lead to one's disqualification. Having a foreign name is a disadvantage; it would be well if one could change his surname to the land in question's common surname. Being married to a citizen, usually a native, could be a big advantage.

Yet the fear of the foreign organism here is less as compared with the fear in the top level jobs in the country in question. People come into contact very easily with the immigrants, and as a result there are many interactions here that make it easier here than the top-level jobs. When we apply the mathematical model of correlation (see Figure 1 & 2) it seems:

Figure 1. Less fear correlates with higher employment rate

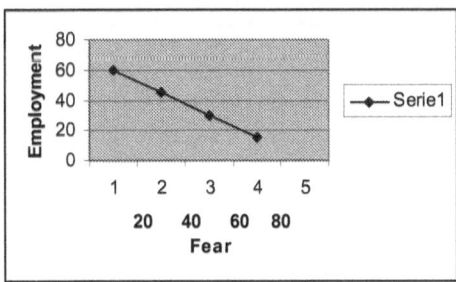

Figure 2. Less fear correlates with less hate

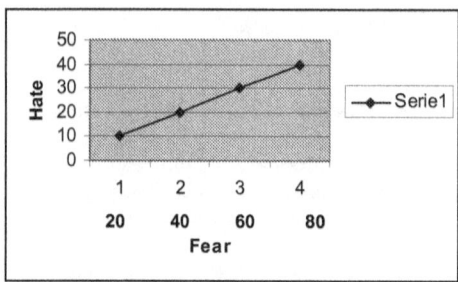

Box 2: The common safety principle

First, less fear for the *subjectuses* correlates with a higher propensity to secure jobs when the soft conditioning incentive is employed. Moreover, less fear correlates with less hate for the *subjectuses*.

COMMENTS:

This situation can already be seen in several societies as many people are employed in the ground level jobs such as cleaning and washing dishes. In some countries, there is no attempt to employ people in the managerial positions even though there are thousands of people with better qualifications and experiences from well-accredited universities. The model says that since the fear of the organism is less here, there are many people with immigrant backgrounds who are employed in the different parts of the countries which support the assertions that less fear gives the possibility of more being employed. These are the ***subjectuses****, a term to be used for the different migrants that are not natives by blood. Though they possess appropriate status in the society. I shall term this type of employment as the* ***common safety principle****. Since without their being absorbed into the job market there will be no peace, and probably many immigrants will turn out to become thieves and steal, or cause injury, a situation that will cause no equilibrium to the native subjects as well as to the foreign organism. In summary, less fear correlates with less hate for the subjectuses in the society. Less fear correlates with higher employment rate for the subjectuses*

Of the Principle of Common Safety

Man is by nature a created being who has to utilise both his mental faculty and his physical strength in order to survive in his environment. It is through work that he is able to fulfil these two important needs; that is, work enables man to exercise his brain and his physical body. Where work is absent, man falls ill or becomes timid and it is a subjection, which only a few intelligent beings on our dear planet could withstand. The *common safety principle* recognises this and makes the fundamental attempt to put the *subjectus* into the job market to enable him to apply his mental faculty and physical body, the activity that generates satisfaction and pleasure.

The common safety principle should be the happiness and greatest challenge of the state, that is, to ensure that the *subjectuses* are *caressed* in a proper manner, which in principle also brings revenue to the state (i.e., through the deduction of income tax). The work of the statesman or legislator that has submerged his mental faculty in the study of political discriminology will ensure that the care of the *subjectuses* brings in returns to the revenue. Another return will be efficiency and better work moral to the working sector. By common safety is meant that all the happiness of the society and security depends on the fact that everyone in the society is gainfully employed. Any idleness on the side of the *subjectuses* will threaten the common peace and security of the nation. It is the interest of the community, yea the whole nation, that persons constituting the *subjectuses* are well taken care of. The interest of the community then is the sum of the interests of the several members who compose it. Since the interest is of great importance, the action or the force impressed by the authority to seek comfortable plans to make the common safety work is paramount. Any short-sighted plans that will be carried out which will not make the principle work must be avoided, since the general tendency is to augment the happiness of the whole nation and not any particular interest group. As Hobbes has pointed out in his great book *Leviathan*, "The *Wealth* and *Riches* of all the members, are the *Strength*; *Salus Populi* (the people's saftey); its *Business*; *Counsellors*, by whom all things needful for it to know are suggested into it, are *Memory*, *Equity* and *Laws* and artificial *Reason* and *Will*; *Concord*, *Health*; ...."[20]

Where the common safety principle is absent or misapplied happiness and satisfaction will diminish, and this will lead to disintegration or chaos. This will be a dangerous consequence for the nation. Shortly discussed below are certain measures that impede the common safety principle.

Different manners of discouraging intake of the *subjectuses*:

Mobbing. When an immigrant succeeds in securing a job in one of these sectors he/she could be mobbed. Only those who are strong and are prepared to continue working, whatever the trouble may be, would remain in their jobs.

Short term employment. Some immigrants receive jobs and while they are being employed, in the beginning it was thought to be permanent. Sudden excuses can be made and this will lead to their being sacked. Excuses can be lack of funds, new structuring to be made, moving to a new building, or "we are nearly bankruptcy."

Criminal background or Imprisonment. Many people in the society have offended one way or another, and may be either sentenced or imprisoned before. But in the case of the immigrant this means your name has been tarnished, you could never secure a job anywhere in the society.

Language problems. Not being able to speak and write the language can be a disadvantage.

Citizenship. Though this is not the case with many jobs, some employers can use this as an excuse.

---

[20] Hobbes (1651: 9).

The implication/Consequences will be:

- *Economic impoverishment*

When the common safety principle is misapplied or not utilised this will affect the financial economy of the *subjectuses* and this will in turn affect his standard of living. Lower standard of living means the entailment of health risk and malnourishment.

- *Lack of security and its resulting lower equilibrium level* to both natives and foreign immigrants

To be in the lower equilibrium level and the ensuing of balance of power will not be good for both natives and the *subjectuses*. It shall increase the crime rate, and this will consequently reduce general happiness or pleasure of the nation.

- *Segregation or difficulties with housing*

Unemployment affects the possibility to live in certain areas of one's choice. The inability to secure and pay a reasonable rent can lead to segregation in the society.

- *Hostilities against natives*

When the *subjectuses* cannot get jobs and there are no proper ways to help them pay their rent and general livelihood, it is probable that they will indulge in hostility against politicians, certain authorities and groups, and this will certainly affect the peaceful atmosphere enjoyed by the whole population.

The *Subjectus* and "Union" Contract

Getting married to a native of the land in question is a good thing. In some countries it not only ensures you, the *subjectus,* the opportunity of getting a stay and a working permit, it gives you the passport and freedom to enter lower and higher places in the land. Some people are able to secure better jobs or positions through this union contract. Others could through this contract purchase a house wherever they want in the corners of the country and reside there. The strategy of taking a wife from among the natives is an old tactic long before the Roman Empire, and is recognised everywhere in the world, though in some countries it does not entail any significance. For attaching yourself holds only so long as the native wants; if this union breaks, you return back to your old status again. I nevertheless consider all marriages that occur between a native and a *subjectus* to be a *convenient union contract* for the simple reason that there is always ulterior motive behind; if not from the native, it will be from the foreigner. Even where the partners may have a genuine motive, a friend who did the introduction or organised the union, a relative or parents may have some hidden motives. For Westerners, in particular, do not want to loose in any game, by their nature they must always win or the rules will be changed in the course of the game to their advantage.

While entering into such a union is a unique manner of obtaining recognition, if the contract is not based on love many strange things could happen during the period when this was at its zenith. For the same reason that you have entered into the union contract because of the permission you are after, the other partner (be it a native or foreigner) may have entered because of other latent purposes. Some of them may be financial others may be slavery or domination. The financial is where the man or the woman wants to get a child with you so that during divorce you could be used as a servant/slave (in Greek, δουλουζ), whereby you pay child support to him/her. It may be considered slavery if one partner has already purchased a house or estate and since you were not originally with him/her at the beginning of the purchase, though you do not own the house, yet because you cohabit with him/her you will have to support him/her for the payment. In such a situation where one is dominating the other it is very common for the union to disintegrate quickly, since it usually lacks genuine trust. And though people usually do not consider it a problem, it ends up being that one of them is used to pay for the other's debt. The most pathetic thing is where one greedy partner poisons the other or murders him/her with the intention of owning the property alone. This has sadly happened to innocent victims who thought at first that their partners loved them and later realised that they were used. Those *subjectuses* who discover such fake love should immediately withdraw from it before something worse happens to them.

It is, therefore, imperative that such union contract is preceded by a careful examination of oneself and his partner, and sometimes the help of an expert is needed before one enters into a union. The fact that there are many problems involved in such a union contract does not mean that people should refrain from it. It is beautiful when partners love each other, and many important men and women have been born into this great world through these "fake" unions.

In many cases it will be appropriate for individuals to discuss it openly what you are after, in other words declare your intentions to each other so that any shortcomings that will crop up later will not provide a shock. If it is an association that is perceived to be above that of convenient union, the individuals should rehearse properly the marriage maxim "till death do us part" several hundred times almost a whole year prior to the union contract. By doing this it will enable each partner to *apply wisdom/reason into the marriage so as to prevent the one-sided "love relationship" that disintegrates quickly as soon as one partner is hurt emotionally or love vanishes abruptly.* The contracting party will realise that the capacity to attach love to reason will make the relationship hold longer, just as friends do when they fight and come together several times. Or when there is a breakage of the relationship no one will be hurt, as it will be considered like a friendship. For man is born free, and even in relationships no one should be made to feel compelled physically or mentally to be in chains.

Figure 3. Union Contract Marriage (We live as friends in marriage when there is no more love)

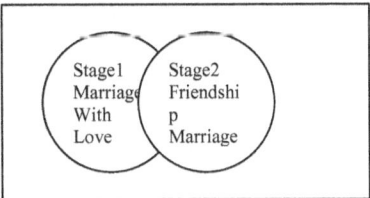

Some Perceptions and Actions of those who Discriminates

Some persons from social institutions and schools of instructions in certain parts of the world think that by allowing only its seeds into the teaching areas or professions, it indicates their uniqueness and "distinctiveness." Furthermore, it accords them respect as against others who allow all sorts of people into their institutions. The question is whether how they perceive themselves coincides with how others visualise them. Recently, a man was standing in front of a kiosk and he retorted to himself that Harvard University was the best university in the world. A young boy who stood beside him heard it and responded immediately that he thinks the best university is not Harvard University. "Why?" the man asked. The boy answered that he thinks the best university in this world will have only white students. People have different perceptions for the same thing, and they may disagree on a simple basic fact or information as to which is the best university in the world.

Solomon was the wisest man the earth had ever known, according to the Jewish History, but his enormous wisdom did not prevent him from acting foolish. Unlike many Ancient Kings, Solomon acquired for himself 300 wives and 700 concubines. A very strong man indeed! The wisest of all men on earth who presented the most "expensive prizes" even to the Queen of Sheba in the Ancient world finally had to admit his folly and in the end he said "remember thy creator in the days of thy youth." Oh, if all people who deceive themselves as wise could realise their folly and apply wisdom while there is much time. The Romans with their most learned culture and well-developed civilisation that opened the eyes of many nations including the Britons, eventually had to endure the most shameful conditions after their empire was conquered, they became the worst of all slaves on earth, according to historians. This is not what everybody/ every nation wants, but in the matter before us we are inclined to assert that we should be each other's keepers and aid one another so that when the evil day should come upon us, as citizens of the earth we may support one another to come out from the depths. Socrates' enormous wisdom could not prevent him from dying from poison. If your conqueror has not arrived you always think you are a superhuman that cannot be conquered. Do not be too naïve all the advancement made is stealing from nature. There is nothing new under the sun!

"The Distinct" or "distinct" individuals and all other people have one levelling formula: "death and the six-feet grave," which as mortal body organs whether from north or south, east or west, green or yellow, we shall disintegrate sooner or later. One philosopher asserted that the equality of man was once to be plainly put to test and be examined, but while he was about to commence that testing suddenly he realised that no one has ever ascended to another outer planet after death, that simply proved his hypothesis that no one is superhuman, we are all cowards, and remain during death in this dark world.

When the historical Jesus was born in Bethlehem, he was looked up by many of the Israelites as the Messiah, the King who was to liberate the Jews from foreign domination. But there were others (the learned, the professors, the Rabbis) who by their prejudice and arrogance prevented them from recognising the "King of Kings" as the Messiah who had come to his own to bring redemption to them. Due to their pride and self-righteousness and their own conception of being superior to other nations, the King came and returned unto his Father again without them recognising the "Greatest Gift" ever given to man on this planet earth. In the end Jesus *reduced* the Jews and *induced* the inhabitants of the whole world into the kingdom, which the Jews had always thought it was only theirs.

Scholium

Since the common safety principle educates us on the need to consider the interest of the majority living in a country, and not a particular group of people, let the nation that thinks about only its citizens be now concerned also about its *subjectuses*. History has a lot of lessons to give to us, that many nations that became powerful and ruled for a long time, after sometime passed on the leadership and civilisation to the next in line. Many centuries ago, the philosophers, those who championed the Enlightenment, noted themselves that though progress and improvement was inescapable, there was also the tendency that it would be temporary. Furthermore, they were convinced that civilisations are cyclical affairs, rising and falling through a life much as individuals do, and that all improvement has to be paid for, somehow and somewhere. After all, the philosopher David Hume had once asserted that "No advantages in this world are pure and unmixed." At present you are up and look super, the next generation or several generations to come you will be down, or probably fighting to maintain the same standard you have enjoyed for a period. It is like that; the world that rotates around our dear sun comes back to the same beginning always. In order for another civilisation to emerge, the existing one usually has go out of the way before the next will come. These historical data are not to threaten anyone, but instead to signal the need to be aware of and to cater for all human beings on an equal basis.

Chapter 4: The Hard Conditioning Incentive

The hard conditioning incentive would be appropriate to be used in certain positions such as Principals, Vice Chancellors, Directors, Company directors, Central Bank Chief, Judges, Government Ministers, University Deans, and so on. If it is necessary, the hard conditioning incentive should be directed at government-owned companies or the public sector before it can then be gradually shifted to other private sectors in the society.

These positions are usually reserved for certain classes of people in the society, and as such very difficult for an immigrant who has adopted citizenship to secure. Moreover, some of these positions demand several years of experience, and because immigrants are not allowed to obtain certain jobs in the society as a whole, then it would certainly be impossible to reach there or obtain the necessary qualifications. It would only take the hard conditioning incentive to allow an immigrant with citizenship to get in, and even if the government should allow it, still there are some enormous obstacles to overcome. In the first place, there are very strong organisations that are sponsored to support the status quo, and these would do their utmost best to discourage any changes that would be tried on them. Apart from some secret organisations that are well paid to protect these positions so that none of the "parasites" could be allowed to come in, still there are the young underground groups who due to their brainwashing, could accomplish any task these people at the top entrust them to do. Secret meetings are held regularly to conduct surveys among those they see as a threat in the society. In some countries it is very difficult to distinguish between these organisations and the government secret agents in the country. Some secret agents of certain countries have dual responsibilities to protect the country and its "insurgents" that want to aim for higher positions in these countries. It cannot be said for sure that the governments of some of these latter countries have knowledge that their secret agents are involved in this venture.

Certain countries have made clear this demarcation and would not let it go easily for an immigrant with citizenship or third generation immigrants to come in and secure top level jobs. Though nowadays what I consider as a form of hard conditioning incentive is allowing more and more of the latter to be part of the top elite. In the UK, the House of Lords can be given as an example, to which still the hard conditioning incentive will need to be applied if some changes have to be made. In the case of the USA, the Skull, which consists of certain elite from Yale University, is now giving way to some generation of immigrants to be part of the successive ruling governments, higher public offices and so on. This is a quick shift from the far right to the centre, which is beneficial for such a land with immense multicultural backgrounds. In Sweden it is becoming more and easier for people of colour to be part of these elite, though it seems some underground groups find it comfortable to see an immigrant with white colour given this position than a person with dark colour. But this is temporary, as within a matter of time and hopefully the hard conditioning incentive, whether employed or not, will lead to reforms in this area in the near future.

Box 3: The dominant principle

Greater fear of the *subjectuses* correlates with greater hate/unwillingness to allow them to take positions in the higher working sector. Greater fear correlates with low employment rate for the *subjectuses* in these top-level jobs.

COMMENTS:

*The interpretation is that the fear of the foreign organism is greater in these higher places and this is understandable due to many reasons that have been discussed below. These positions are sensitive and they consist of men who want to dominate and keep things under their control. It is no wonder, due to the greater fear, the possibility of getting in is very slim for the subjectuses. That is, the greater fear causes greater hate or fear to be absorbed in this level of employment. Nevertheless, I shall term these levels the **Sovereign**, those that have obtained these positions through the **Acquisition or Dominant principle**. Their stronghold to power makes them use hidden iron hands on any threat that might arise from the minorities. The hard conditioning incentive will be able to lessen such a problematic situation. In summary, greater fear correlates with greater hate for the subjectuses. Greater fear correlates with low employment rate for the subjectuses*

There are some myths and false ideologies that are cherished by some authorities in some countries that when they allow immigrants with citizenship into their universities as leaders or principals, companies' directors, it will tarnish the reputation of the universities. This is a myth that needs to be conditioned and abandoned. History has shown that many Jewish immigrants who were put in important positions in universities in some well-known European countries before World War I attracted students when they were advertised for their expertise. The fact that some other person with different colour can do the job better than the native should not prevent any authority to maintain negative attitude to people who could equally do it better (cf. figure 4 and 5.).

Figure 4. Greater fear correlates with low employment rate

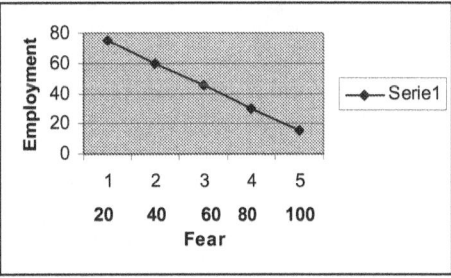

Figure 5. Greater fear correlates with greater hate

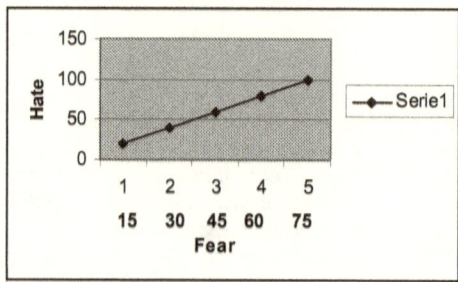

The Dominant principle

I have employed this *dominant principle* first to imply that the dominating ruling class that entails the sovereigns in principle dictate who should be in these positions; that is, either through voting or through personal election by the same dominant figures in the society. Secondly, the dominant principle also depicts the greater amount of influence the sovereigns have in the community or the nation. In every nation there is supposed to be this group of persons that have power and rule, and their authority in political terms is said to be Divine, or sent from above. In other words, their power is divine and it is the ultimate power in the land. The sovereign is said to possess the sovereign power, and everybody else are their subjects. Their authority which, if it is kingship, may be said to be acquired. But in representative governments these people are elected by the people to represent them or become their mouthpiece.

Since in political matters years of experience are very important, not many *subjectuses* can have the opportunity to elect or can reach this level, as the requirement is high. Therefore attaining this sovereign power can be done in two ways. First by the dominant principle, where the hard conditioning incentive will be applied to augment the number of sovereigns in the society in question. The other is using the puppet sovereign to fill some of the positions that have become vacant. The latter is always viewed with suspicion, and therefore it is not the ideal type of thing to do. But when the super-ordinate force (government) applies the quota-merit principle, then it will remove this idea of putting favourites there, since it jeopardises the relationship between the sovereign and the *subjectuses*.

The Sovereign: the puppet and the forfeited

Through the dominant principle the acquisition of some positions in the ruling class is made possible to the foreign organism. It is through perseverance and hard industry and patience that people are able to get the opportunity to fill certain positions which become available not only through the help of the natives, but also through the voting contest. The sovereign status to the foreign organism must be protected, because the underground organisations usually vow not to sleep until they have succeeded trapping the former through different means. Some of the trappings are fixed with care that only the clever person would be able to outwit. These can be presented to you through a woman who may attempt to seduce the sovereign into bed, and then later be accused of a crime and falsely charged. It can be presented to you in a bribe form, and then you can be charged with receiving a bribe. It is also possible to use a member of staff in a lower rank that may commit an offence in one's jurisdiction, and the sovereign's inability to rectify this may cause him to be removed as incompetent. In short, there will be a series of temptations with the purpose of finding you committed a crime or a major mistake that can give the occasion for you to be removed. It is here that if a sovereign is married to a native, she/he will be of help, for since they know all these tactics their people utilise to cause the downfall or to undermine people, she/he will be of help to advise you concerning these vices. In some cases, the fact that you have attached yourself to "one of them," it may be possible to prevent them from doing anything that might harm your reputation. A sovereign who through these successive plans and sabotages looses his/her positions becomes the *position-forfeited sovereign,* and in many developed countries a sizeable number of people have lost their positions through this manner of sabotaging the foreign organism.

The *puppet sovereign* usually has favourites among the natives and he has been offered a position because of his *good-for-nothing personality*. Some of these people are unlettered, and they will not even object to the taking away of their last clothes they wear if requested by some of the people who helped placed him/her there. Such people, when chosen to represent a group of people in, say, organisations or the parliament, may spend infinite time there without people recognising what contribution they have made. Because of the manner they became elected they are compelled to keep their mouth shut like the lamb, which Abraham offered during the sacrifice on Mount Moriah. Because of their "fixed" position, only the people that forged to put them there cherish them and they are people that the *subjectuses* detest, though the latter pretend that they like them because they have no choice of removing them. Some of these men or women become corrupt, since absolute power corrupts absolutely.

The *straight sovereign* is usually critical with the system, and the criticisms that he has for the system is not without cause, he wants the best for his adopted country and so he does his utmost best to see how he can contribute positively to the society. Despite his occasional criticisms of the system, nevertheless he has deep respect for the leadership and co-operates with them on many important issues affecting the nation.

I shall now present my objections to the uprising of racism, which some people who oppose this principle believe the utilisation of these common incentives will bring about. Still, the reader should understand that discrimination is a threat to the modern society, and as this is a theory building, we should not be surprised if new ideas emerge that will call for the need to incorporate into it.

Chapter 5: Objections to the Uprising of Racism

Racism is defined as a belief in the superiority of a particular race; it is prejudice based on this view. In a radical way, racism is the antagonism towards (as a rule it goes also with the reverse, antagonism towards oneself), or discrimination against, other races, especially as a result of this fundamentalist belief in special race superiority over others. Broadly speaking it is the theory that human beings' abilities and so on are determined by race. The individual who submerges his mental faculty in these beliefs and sets out to practice it to the letter is referred to as a racist. In our present generation this belief has caused some people to do untold harm ***first to themselves, secondly to the world,*** and ***thirdly, to other different races scattered all over the planet.*** Let me first demonstrate briefly each of these propositions to which I have termed the three predictions of racism.

The Three Predictions of Racism

Racism Causes Harm to Oneself

To be a racist one certainly causes harm to oneself, in that the hate one possesses against someone may have its repercussions. The psychologist may say that the amount of energy that one invests in this hate usually drain the doer so much that it could lead to disappointment and the consequences of depression, etc., if one is unable to carry out the cherished deed to the person one hates. The racist idea breeds militancy, and is so dynamic that in some countries racist ideologies compel some people to take guns, or any weapons which can results in the death of the person one hates. The result is, if one is caught in a criminal act that involves racism it will lead to imprisonment, and suppose the person got away with the crime without being punished, the guilt and its associated symptoms may lead some weak individuals to deep depression, that to those that are not properly protected, could lead to suicide.

Racism Causes Harm to the Physical World

People are beginning to realise how small this world we live in has become; in fact, now the world is like a global village. The anger and hate which makes people take weapons against individuals, groups, or a nation usually result in some damages to the environment. Racist wars have left some wounds and scars all over the planet. These wars have brought enmity between nations both big and small, as well as caused destruction to famous houses, expensive factories, great church buildings, important monuments, great libraries, paintings, etc. It was after World War II that the world and its inhabitants discovered what Germany did to the environment in Auschwitz, and even the destruction of the city of Berlin was one classic example of how the hate of another nation could lead to mass destruction in the world. The recent incident in the USA, which sparked the people of the worlds' attention, is also another example. Timothy McVey was a young man who shocked the world by bombing the big apartment building in Oklahoma, for the simple reason that he believes the Federal Government is impinging on their groups' freedom to use weapons to protect themselves. He was a member of a racist group that had weapons to protect themselves against the Federal government.

Racism Causes Harm to the Feelings and Integrity of Other Races

The use of the mass media such as the newspapers, television, and mobile telephones to report certain incidents involving racism call the attention of the world to it. The world is a global village, where today nothing escapes the interests of the inhabitants of the world. When racism goes on live it is reported and the people from the different races in all corners of the earth see it. There are sides taken, which can be very strongly opposed or supported by other people of the same race physically. Nowadays, sanctions from countries where the majority of the race live can lead to instability. Other races take sides, and it may gradually lead to religious wars or holy wars. The Australian Nation had to put an end to its policy regarding the discrimination of other races when the USA put pressure upon them.

From the discussion of the use of these two common conditioning incentives, that is, the soft and hard incentive conditioning, to help the foreign organism to sojourn in a land where he had adopted citizenship, I have made it plain why it is necessary and appropriate to make the *common safety principle* of employment available to the *subjectuses*. From the point of view of the theorist, it was also essential that the *dominant principle* allow the hard incentive conditioning be used to enable some foreign organisms to be part of the *sovereign*, those that form the strong ruling class, since their role in the society also has to reflect the power and strength of the society. It is a common saying among sceptics and other people that the employment of these conditioning incentives would generate a kind of hostility that might result in racism. This is what I intend to argue, and to submit some points that I think would cast shadows upon the worry and apprehension that the situation will degenerate to envy, hatred, or war.

The Objections

First, the principle of representativity that the theory employs is a chief democratic concept whereby the *subjectuses* are represented in all areas of the society. This follows the attempt to have equal distribution of income that most governments make as its primary aim when they are voted to power.

Secondly, the submission of the will, right, and power to the governing body of the state by the *subjectuses* means that the former fulfil its promise to the latter, that while governing it will protect them from being maltreated or mobbed. It is a breach of contract if the *subjectuses* are not able to come into the job market and compete fairly and squarely.

Thirdly, the *common safety principle* of employment and the *dominant principle* all follow the natural laws of justice, modesty, equity, and mercy. The skill of maintaining common-wealth, according to Thomas Hobbes, consists in certain rules, as do Arithmetic and Geometry. The laws of nature are eternal and good, and when applied correctly, it ensures peace in a democratic society.

Fourthly, citizens have no course to be afraid with the employment of these incentives if they believe that there is only *one world* that is habitable at the present time, and that human beings breathe the same air and possess mortal organic bodies which sooner or later disintegrate. For injustice, ingratitude, arrogance, pride, iniquity, and sorting out persons according to colour is not *natural*. These do not preserve peace; they destroy peace and life.

Fifthly, it is high time the government took a serious look into those who discriminate and possess what I have termed in my theory as *resistances*, (i.e., myths, false ideologies, fables and personal prejudices). Some of these people need to be helped, as their mental ideas and slight possession, which they reveal in the course of telling their experiences when caught in crime, depicts some kind of mental disturbances according to some experts.

Sixthly, history has shown that in the Ancient Kingdoms work or labour was the only thing that kept citizens, as well as the immigrant subjects, as part of the progressive society. Continuous labour in building infrastructure kept citizens as well as the immigrant subjects (the conquered, the slaves) away from crime and insurrection. Where the King or master worked and the servants sat and were adored, it was thought to be an unnatural order of things. Something was considered being wrong!

Seventhly, and I say any man *whosoever persecute* its *subjectuses* (foreign organism) not for any other reason but for wanting to become part of the *Sovereign* through the *dominant principle,* will lead itself and his people to extinction. You should cast your eyes to such nearby countries now and of old that had attempted to do that by subjecting their *subjectuses* (foreign organism) to the consequences these people suffered. In the same manner, whosoever encourages its *subjectuses* into becoming part of the *Sovereign* would save his land and his own people from doom. The case of Joseph in Egypt (the only *subjectus* who could interpret the dreams of the King of Egypt) who became king and saved his host country from seven long years of famine; and Daniel, who became the only person to have been able to read and interpret the handwriting on the wall in Ancient Persia. Winston Churchill, who became the leader that with the help of the Allied forces fought and defeated the leader of the Land that started World War II.

Lastly, it has been observed that discrimination is the problem of the human organism where even the well-educated person in a most cosy famous university somewhere in the developed world, or the illiterate bushman in a remote part of a continent, has difficulty in dealing with it. And judging by the fact in future there is going to be many problems in dealing with people moving across different boders in search of a better life in the developed world; and in the case of the citizen of the developed world, people will run away for shelter in case of a major natural catastrophe, like for example the sinking of a European Island or an Asian Island, a major volcano erupting in the middle part of Italy or the USA. And seeing that there is only one world among the many planets that can be inhabited at present and the fact that no one person or a nation can claim part of the world absolutely as his or theirs respectively. Because if we do that, the Saamis will claim Sweden as their own land; the American Indians will claim the North America as theirs; the Amazon Indians will claim South America as their own land; the Aborigines will claim Australia as their land; the Tasmanians will claim New Zealand as their own land; the Zulu will take South Africa as their land; the unknown Islanders will claim the British Isles as theirs; and the Cannanites will claim the Palestine as their own land. And understanding that the Great God intended it to be that we all live on the same *one* planet earth, and the fact that even certain living creatures like Ants and Bees live sociably one with another which Aristotle called them the most political creatures. In connection with my *Common safety principle* that seeks means whereby people can live with one another irrespective of the social background of being rich or poor, *I propose that the study of human organism in the matter of how to live with one another on this one planet earth; (which has been ours in totality from the beginning, but has been divided because of nationalism some generation of years ago,) be called DISCRIMINOLOGY.* This will study the "distinction" (in Latin *discrimen –minis*) that has come about because of the environment where human beings lived for many thousand of years. For I personally do believe in inborn excellence of all God-created beings, but the environment has caused the differences in the way people think and solve their daily problems. The contribution will be to discern those factors that have caused the organism to be different (e.g., the extreme sunshine in certain parts of the world, and the three elements that it sends that cause damage to natives living there, or how the extremely cold weather causes weak bones and certain incurable illness in the poles), and how the causes of the deficiencies can be eliminated for good or fought against. All with the supreme purpose of bringing human beings to accept one another without trying to sort out human beings in a negative way, and to preserve what the immortal God has ordained it to be. It will be a major step in preparation, for any surprise flees from the so-called developed world (in case of major natural catastrophe) to the undeveloped world, as it happened so many million years ago.

Scholium

As many states both old and new, that is, underdeveloped and developed democratic states, are becoming more and more aware of the need of depending on one another, the suggestion to have a separate scientific discipline where one studies the science of living together is imperative. *Discriminology, therefore, is the new branch of science that investigates the manner in which human beings live together as one people on the planet earth.* As a separate branch, this new discipline will not only base its research on contemporary issues affecting the supreme created being on earth, but it will probe the human race as a unique created body that is competent to deal with the differences that exist between *mannian* (man, from Germanic). No longer shall we go to war based on the simple fact that one nation has offended the other, but that we shall be very cautious regarding any misunderstanding that crops up between the human race, and be courteous to one another. The difference between this new scientific branch of discriminology and most other human sciences will be the manner the former will base its investigations on, using rigorous methods to make the human organism able to find ways and means of how to live together on this planet without much friction. The new science will incorporate the study of astronomy to get a better view of the universe where human organisms live, and to understand man as a unique creative body or creature among other big bodies in our solar system. It will focus on the natural laws that govern both the human organism and the bodies in our system to allow us comprehend how man can live harmoniously on this planet, and extra planets we intend to inhabit in the future.

With one mind and purpose we shall regard ourselves just as the five figures we have on an arm, each different in height but also able to perform a specific function better than the other. As each recognises its specific strength and duty but still depend on one another's role, so also shall human beings come together in a unique manner as one and work together on how to live as one *body* or people. These topics should be the concerns of the new philosophy: Liberty and freedom for the organism; respects for Human right; respects for the natural laws; the Great Author in a new perspective; reason, understanding and science; and understanding our solar system, their unique state and orderliness.

Chapter 6: The Theory of Superiority Complex: Personal and Collective Experience

To try to write a grand cosmical drama leads necessarily to myth. To try to let knowledge substitute ignorance in increasingly larger regions of space and time is science.

Hannes Alfvén, Nobel Laureate, 1970

In my opinion, a theory of superiority complex will be able to illuminate on the general picture of what prevails in the society where most *subjectuses* find it difficult to actualise their potential. My purpose here is to propose a general theory of superiority complex that will be capable of specifying certain principles, which will explain the lack of the appropriate equilibrium in the working sector, which will also have strong predictive power. In our attempt to provide a vivid description of the theory, we shall introduce radical innovative concepts such as equilibrium, actualisation of potentialities, irreversibility, self-sufficiency, recognition, and a few others. The proposals in this theory are very dynamic because they point out the necessity of self-searching to see whether a person's normal self-perception is that of genuine nature or something that suggests abnormality in behaviour. In conclusion, these conceptual innovations of superiority complex are added to the theory of selfishness that I try to propose at the same time, with the support of strong evidence and embellishment, in order to suggest the necessity of utilising the problem-solving theory of non-discrimination law.

My design in this chapter is not to explain the development of this complex by hypotheses, but to propose and prove them by reason, observation, and if necessary, by experiments; in order to which I shall premise the following definitions, and then later with the axioms. For as have been indicated already that a real definition may always serve as the premise, or part of the premise, of a logical inquiry concerning a subject matter. Thus from the definition of similar figures, together with our premises we can deduce our theorems which will follow later.

Equilibrium Concept

Equilibrium comes from the Latin word *equi libra,* meaning 'balance.' In the matter before us that I want to expatiate on, the word equilibrium means a state in which the energy in a system is evenly distributed and forces, influences, etc., balance each other. Usually such a description fits well the state where Arms races between countries come to a point of a stable equilibrium; that is, the state of equilibrium arms level indicates a reduction of tension between those countries that are competing to surpass each other or aiming for supremacy.

This is a stage where both countries feel satisfied with their armament and do not feel threatened any longer. In connection with our theory, one can speak of *higher equilibrium* and *lower equilibrium states* of the *subjectuses* in the society in question. The latter, that is, the lower equilibrium, is where the *subjectuses* (foreign organism) through the common safety principle have managed to secure a sizeable percentage of the labour force in the market, mostly in the lower sector jobs such as cleaning, dish washing, work in the pizzeria, store assisting, news-paper distribution, and a host of others. We may regard these areas important for the economy of many of these individuals because their employment, to be understood, affects not only their economy but also their survival in the system. Though the income from these jobs are not high yet, judging by their original backgrounds and the countries of origin, they are able to make ends meet and furthermore remit some money to their extended families in their home country. The low equilibrium means that there is no balance in terms of their influences or forces, and the general situation of the *subjectuses* in the country in question become very pathetic, though not to the extent that they go about being famished and not able to cater well for their expenditures. Many immigrants in certain countries where they do not obtain regular employment find themselves in dire situations. Their situation in terms of making ends meets depends on the social help they receive from the Municipal authorities in the communities in which they reside. These financial institutions are well organised, and they include financial help that enable many families and their children to live nicely, as well as raise their standard of living to the minimum level. When this situation occurs, the individuals may experience the ***inert state*** of their career where their potentials may not be adequately used in the society. Though full of energy, those that cannot cope with this inert state may turn to drugs and other criminal ways for survival. The inert state that denotes a state of full energy yet not capable of being utilised is frequently experienced among the societies that possess superiority complex. We shall discuss this below in detail.

      The higher equilibrium level is seen in the certain states where many *subjectuses* have the greater chance of being employed in the higher position levels through the common safety principle. Here, *subjectuses* have equal opportunity of being employed in the managerial positions. Where they have this possibility of making decisions in terms of themselves, conducting interviews for many employees, and also making top decisions, it raises the low equilibrium state to higher equilibrium state. This means that there are unlimited possibilities for the *subjectuses* in the system that helps them to organise well, as well as function properly in the country in question, though not to a level that may indicate that they have acquired *sovereign* status. For to reach the sovereign positions requires a special strategy that I shall discuss below. The distinguishing factor of the higher equilibrium state from the low equilibrium state is the ease in which the *subjectuses* who have reached the former state conduct themselves in terms of being more industrious in the setting up of companies, and employing more foreigners/immigrants that increase their influence in the society. The society becomes the *subjectuses* and the *subjectuses* become the society, where they do not complain of a problem in acquiring work. If the low equilibrium makes the *subjectuses* lie in the horizontal level in terms of their complete general economic situation, the higher equilibrium makes the *subjectuses* stand on the vertical level. Equilibrium must be understood to be more a state of having positive influence in the society and the ease of getting what one wants in an acclimatised system without being pushed around. It does open many doors, and becomes a bridge where people with luck can walk on to the sovereign status. The higher equilibrium state may be compared to the ***kinetic state*** of the human organism where adequate use of his potentials in society is felt. Kinetic state is where the individual recognises the full utilisation of his potentials or abilities in the society in question. At this state, in fact, one can speak of ***actualisation of a potentiality*** of the individual in question. Thus kinetic state contains the zenith utilisation of the individual's potentials. The dynamic development in the kinetic state presupposes that the individual has superposition-inducing dynamic process complementary to the sovereign status.

Equilibrium, Democratic Tradition and Complex

Democracy is defined (late Latin *democratia,* from Greek *dêmokratia*, that originates from *demos,* 'the people' plus CRACY) as a system of government by the whole population, usually through elected representatives. With this system of government a state can be governed through the holding of an election (3 to 4 years intervals) to choose its representatives and leaders. An organisation or a state that is governed by this principle of governing can be referred to as democratic. The employment of this democratic principle to govern is popular in Europe, both Western and Eastern, as the fall of Berlin Wall has enabled many authoritarian governments to be replaced by democratic governments.

It has been observed that certain forces that exist in the tradition and culture of the people in question also impede the application of pure democratic principles as a manner of governing. And so, while the use of democratic principles may be employed, say, in some secluded country in the new emerging states of Eastern Europe and also somewhere else in the continent of Africa, the appearances of these governments will surely not be the same. In each of these countries one will find mingled with these universal principles of democracy certain traditions and customs that either make it different, or sometimes complicate matters, for the scholar that is not well informed of these cultures in question. Now, I do not want to imply that such manners of adapting democracy to the practices of the culture in question is bad, but on the contrary, I venture to suggest that any scholar of this field will not observe the principles being applied the same everywhere. In a country where these forces of culture tend to dominate the principles of democracy, in my opinion, there is a tendency to allow democracy to remain backward with some traditional and cultural principles mixed in that make it difficult not only for its citizens, but also aliens that immigrate into such countries.

Some of the difficulties that such nations present to its citizens and its aliens are revealed in the manner they treat people. In a country where the people feel some sort of superiority because of their overall general history and experiences with the world around them, the concept equilibrium and power balance is manifested differently among them. If they believe that they have a superior culture, then this will make them arrogant, rude, and there will be a tendency that will make them feel as if they are the best in the world. In fact, this kind of illusion comes usually not without a cause, for they may not have been oppressed before as a nation, or they may not have been conquered. In this case, this may suggest that they have not been to war for many hundred of years. This kind of arrogant attitude may be coupled with the fact that they may have been living in isolation for many years (village community), in which case they had succeeded in maintaining a uniform culture that had not been tampered with by any foreign culture for years. They could also have emerged from poverty or famine, in which case they hold *dear* to them anything they have acquired for so many years. The *subjectuses* who migrate into such a country or society (if ever there is one) will also find themselves nailed down to the lower equilibrium level. The experience of the *subjectus* of not being able to move up to the better, higher equilibrium level, hard as one tries, in such societies makes me term this typical condition *hewers of wood monopoly*. In such societies the *subjectuses* are more or less like servants, in that there is a slim opportunity for them to attain managerial positions or *dominant positions* in the country. I use here the Greek word $θητες$ meaning domestic servant, implying those that serve voluntarily or for hire, or in hope of benefit from their masters. This is different from $δουλοι$ which means slaves or servants as well. They clean and run errands, such as delivering pizza to a customer, or function as store assistants. All the laws and the principles of governing are geared towards this idea of putting the *subjectuses* down at the bottom of the ladder. The use of the word monopoly here means both the inward and outward acceptance by the *subjectuses* to remain in the lower level while monopolising the low status jobs, without making any tangible effort to climb the ladder and claim their **right** for higher status jobs. Their timid behaviour is seen with the way they accept their lot and enjoy the invincible oppression, which the system indirectly subjects them to. There is a kind of fear disposition, which is inversely revealed in the manner the inhabitants of the country in question pretend to love the *subjectuses*. In such societies, they see beauty only in themselves, like the narcissistic that died (before he could be healed) because he looked through the mirror and could find his image more beautiful than any other person does. In such societies in fact, the customs are more respected or revered, and so also are the laws, though because of fear for punishment people pretend to observe or keep the laws. I assign this special attitude of revealing fearful character toward the *subjectuses* yet in a manner of feeling or possessing arrogant disposition as the ***superiority complex***. The opposite is ***inferiority complex***, which is experienced in societies that have experienced oppression from other nations or societies before. In the latter society, the *subjectuses* are usually chased or hunted at to flee from among them (native subjects), though this may be carried out in a gentle manner without attracting the attention of the general public. But when the militant Underground groups engineer this kind of hostility, animalistic instinct is revealed. Those private individuals in these societies that become obsessed with this manner of hostility toward the *subjectuses* can be diagnosed as suffering from ***psycho-superiority complex*** (Psycho-S. Complex) and ***psycho-inferiority complex*** (Psycho-I. Complex) respectively.

But there are some societies that make their democracy superior and their culture second, though such a perfect one may not exist. Usually such societies consist of more different people without being too homogenous. Though the general tendency is to elevate the principles of democracy, nevertheless their culture is important, and the citizens as well as the alien feel this. These countries may make one feel that to become part of them means one has indeed acquired a status that makes one equal even to the prime minister, president, or the prince of the land. They are countries that always look for better ways to make the *subjectuses* achieve its dream. All the measures are that when you are pushed around, boy gets to the right source and challenge! And indeed the laws have no respect for persons, which is also seen in the former case. Yet here the *subjectuses* are the society, and the society exists because they are there. Though there are, of course, imaginary classes of people, these do not in any way make one subject to any propelling force from above pushing the *subjectuses* down. There is the recognition of the common suffrage. There is a tendency to put weight on and learn from past mistakes, wars, nation's mistakes, human abilities, and the pace of human as well as race developments. Such countries that are seen as more progressive, and recognise human abilities and *waste no human resources*, I hereby term them the *good gesture right*. By good gesture, I mean these countries recognise that everyone has the right to become what he wants, not being pushed around by certain individuals that exist in the underground who always want to maintain their selfish interests. In such societies, say, if an individual through nature has been endowed with potentials or certain talents, no one tries to impede him. In short, the good gesture right makes people obtain what they aim at in life with proper care; and with a little ambition it ensures the balance of power that is needed to reach (the key) the higher equilibrium level.

Like the former society, there are some deficiencies in the societies that the *good gesture right* is practised. The conditions that qualify people to experience *superiority complex* and *inferiority complex* exist in these societies as well. That there are numerous individuals who we can say are obsessed and do engage in hostility against the *subjectuses*, and even a large population of the natives themselves, cannot be disputed. These fellows can be easily diagnosed as suffering from *psycho-superiority complex* (e.g., some Supremacy groups elsewhere in the world) and *psycho-inferiority complex* (e.g., other religious sects elsewhere in the world). The rate at which these obsessed individuals conduct their campaigns around both in seclusion and in public cannot be underestimated. These hostilities depict how substantial this kind of illness has gotten its roots in modern societies, and since they are being funded by some well-organised groups in the rich countries of the world, it appears the fight against them will not be easily won. There too, the prisons are the *devil's garden,* where both some natives and the *subjectuses* spend their time to grow and learn more trade concerning criminality. This unpredictable behaviour that originates worse social conditions has given the occasion for the authorities in some of these advanced countries to take an unpopular yet injustice approach to use the prison to keep control and domination over the minorities who are always victims of crime and cruelty. The overall general situation has consequently brought fear into the public, such that one usually becomes afraid and entertains the *threat of shadow* atmosphere in these well-developed societies in the world.

I have given this description so that it may appear to the reader that it is easy to see if a country one lives in fits into these two groups. I should, however, assert that you should not be disappointed if you are not able to do so, because this is only the beginning of this theory building. The conditions enumerated appear to suggest that certain countries cannot fall within these two groups. We shall see how we can modify these conditions or try to include other perspectives.

Complex and Equilibrium

In psychology, complex is a related group of usual repressed feelings or thoughts which cause abnormal behaviour or mental states. But here in our treatise it could appropriately be referred to as a preoccupation or obsession with "something." One can possess complex about punctuality, superiority, etc. The problematic condition that gives me the course to designate a group or an individual as "complex" may not be necessarily interpreted as "illness," so far as the condition has not deteriorated to the psycho-level. I say this, as many of the victims holding these false notions have been wrongly exposed to the teachings unknowingly that make them cherish attitudes that occasion this miserable condition. If one should question someone about it, he/she would respond in a normal way that this is certainly true, and would find nothing wrong in cherishing these false ideologies. In other words some people cherish these notions out of ignorance. The question that naturally follows is that, how have these individuals come to possess these false beliefs that have conditioned their behaviour to conform to the ideologies that subsequently usher many ignorant people into this mess? The majority of people have received these ideologies through early indoctrination at home and in schools in a manner that could never have been rejected by the individual, as they were poured into the mind in an ingenuous way. Others have imbibed these ideas during church services that take the form of readings and singing. In fact, some sermons preached in a Christian congregation have the tendency of propagating these ideas indirectly to many innocent individuals. Group pressure at school and in certain communities exposes many young men and women to these false ideas in the society in general. Another important source of these ideologies can be ascribed to certain organisations that have been set up purposely to propagate these false ideas, and always see to it that they are watered regularly in order to yield harvest. The danger caused by these organisations is enormous, and since they have the resources because they are in responsible positions, they are capable of sustaining a greater damage to both the society in question and its young men and women. Young men and women that are employed by these organisations receive rewards in the form of reinforcement, and they are successful simply because they are able to muster those individuals who have been neglected by the society, or suffer deprivation of some kind. The leaders of these organisations, through emotional speeches and sermons, help trigger these deprivation experiences and some childhood difficulties/disturbances into consciousness due to the preaching of these false ideas into their minds. This accounts for why certain individuals will sometimes eagerly plunge into evil activities that at other times leave them indifferent, and for their remaining restless mobile until certain "goals" or "purpose" are accomplished. Among other things, the measure of ***degree-of-complex*** will surely increase in strength with particular individuals:

- *Superiority complex increases with age among the young adulths especially from teenage onwards*
- *Superiority complex is sustained among the majority of intellectuals who want to remain distinct*
- *Superiority complex is revealed among certain old age males who hail from aristocratic homes*
- *Superiority complex is commonly found among certain church denominations who strongly believe in the false theory of a Biblical curse*
- *Superiority complex occupies at least seventy-five percent of the Underground group individuals' thoughts.*

The ***severity of the effects*** of the complex that the obsessed psycho individuals possessed can cause certain individuals to commit murder or go about and destroy or burn down religious buildings of certain churches. It can drive some obsessed individuals to burn down immigrants' camps and quarters, as it happened recently in France. Some well-respected intellectuals can be obsessed such that they will follow their students they are at logger heads with to anywhere they look for jobs and sabotage these students so that they will be refused employment. Since there are no observable symptoms except these highly charged (*minutrons*) mental ideas and *avoidance*, the hate can never be reduced until the death of the sufferer. These individuals detest any plans the government lays down to foster *integration* of immigrants. They are fond of *inducing conflicts* in certain familiar places in the city. What they detest most is the common news that human beings are *equal and worth the same*.

Complex madness was behind many world wars that were fought during the inception of man's development. And even presently in our modern day, complex madness is the motivation behind unpopular global and regional wars that are taking place between stronger and weaker nations. This shows that the theory of superiority complex's predictive power even goes way behind these great centuries of developments. It was probably this madness that motivated Cain to kill his only brother Abel in the then lonely planet, according to the Biblical myth. And so also as in the great history of the Ancient Babylonian Empire when King Nebuchadnezzar became motivated through complex madness to wage brutal wars against the Jews and other neighbouring nations.

Linearity and Non-linearity of Complex

By non-linearity we mean the process whereby the superiority complex individual recedes to a state where he/she no longer cherishes those highly charged ideas (*minutrons*) of being superior to others. This implies that the complex effect which goes on in the mind has return from a disorder to order, to which we shall call ***superfluity*** for her uncharged mental ideas.

Conversely, the linear complex is the process where the superiority individual makes a lifetime bond with an organisation that makes him/her maintain these false ideas. In connection with her everlasting disorder that results from her mentally charged ideas (*minutrons*), we shall term it ***pinnacleconductivity***. The linearity and non-linearity of complex is determined by four factors: (1) *union contract with a subjectus*, (2) *change of a profession*, (3) *Philosophical-intuition*, and (4) a *lifetime bond with an underground organisation*. The non-linear occurs in the first three; that is, union contract with a *subjectus*, a change of a profession, and Philosophical-intuition of the individual.

In a union contract marriage, individuals that used to be members that are engrossed with these false ideas suddenly make a turn that indicates their relinquishing of those ideas either imbibed during childhood or acquired through teenage influences. The union contract with a *subjectus* means that a person accepts the universal brotherhood of man, and then becomes a champion for the unity of all races.

The superiority complex individual who becomes a priest will also make up by preaching against these ideas. In fact, this person's sudden conversion from these ideas could make a very big impact if, say, he is assigned as a pastor in a congregation that is situated in a community with inhomogeneous members. Some of the people who do not become priests or pastors could go abroad as volunteers through the help of missionary organisations. Usually, while a strongly opposed person of intermarriages, this individual would either become married or allow a member of his family to do that.

A superiority complex individual who through philosophical analysis discovers the truth about the equality of mankind will suddenly give up these false ideologies. He will then devote his time to learn more about this truth. He will later become an anthropologist to help him/her learn more about human beings and their social organisation, or become a social worker or a nurse, who will care for human beings. This assertion has received, in fact, empirical proof because one hardly finds a person who possesses one of these professions acting different to show that she/he feels superior to another person (see Figure 7).

Finally, it can be asserted that there are some people that do not change from this condition in their entire lifetime until they pass away or cease to exist. These are the people that become eternally attached to the racist organisations, and their continuing contact and interaction make it impossible for them to alter their mentally-charged false ideas. Since these ideas are watered and cared for they indeed become engrossed such that not only are they themselves eternally doomed but also their offspring become members of these organisations that have strategies of maintaining the number of members they have, as well as winning more members into their camp. These *external constraints* are the basic mechanisms that compel them to remain psycho-S complex or psycho-I complex, and their overall purpose in life is to see that appropriate equilibrium level is not reached by the ***action*** of the government. The concept ***irreversibility*** implies that the racist individual who has made a lifetime bond or attachment with the underground organisation has no chance of receding from these false ideas that make him/her possess superiority complex. As far as this person is concerned, the probability of his irreversibility is closer to one ($p \leq 1$). When plotted on a graph, the equation (**linear equation:** χ is a linear function of γ, that is $\gamma = \alpha\chi + \beta$, where α and β are parameters that remain constant in the given situation) between the two variables of superiority complex and age will certainly give a straight line (see below graphs, Figure 6).

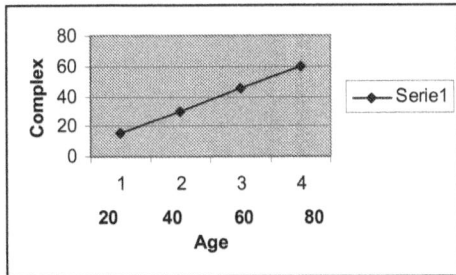

Figure 6. Linear graph for the variables complex and age

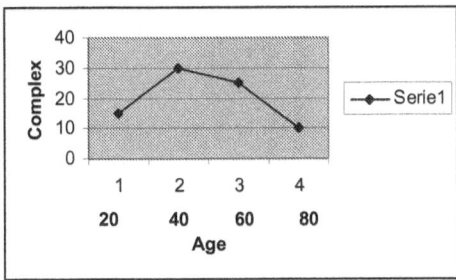

Figure 7. Linear graph for variables complex and age (non-linear)

Pinnacleconductivity and superfluity

The behaviour of most people who become perpetually attached to certain underground organisations and religious organisations that propagate and spread these mentally charged ideas (*minutrons*) are described to be unpredictable. Especially with the way they cause destruction to the lives of *subjectuses* and their properties. Those that are found to be brought up by parents that have affinity with these groups seem to be automatically drawn to these organisations because of their parents' contacts. They might have had prior contacts before the birth of the individual, so that these contacts are mediated to the individual who found it easy to continue this old bond of her parents with the particular organisation. It has been established that those who continue to be members of these religious groups, as well as these secret organisations that have regular meetings, grow well in their beliefs and bonds. There is absolute no way whereby this individual can sever his membership from these organisations unless he happens to meet a partner that does not want to belong to them. Even that would not be an easy venture if one wants to withdraw. The complex individual has no way of discovering his mental state, whether they are normal or abnormal. As already mentioned, the absence of observable physical symptoms, with the exception of avoidance, makes it impossible for members and even outsiders to fairly judge the serious situations of individuals.

This simple picture of this kind of defect mental state makes us rationalise that there may be other underlying symptoms which can make this generate to a serious illness that can call for the attention of experts. Where the individual has had a serious depressive illness, paranoid ideas, or schizophrenic disorder, it can make the superiority complex worsen. As it is not my intention to investigate these probable underlying symptoms that act as a catalyst because of insufficient knowledge, I shall restrict myself to only this condition.

The inability of the individual to get this behaviour corrected is worsened by the continuous preaching, indoctrination, and brainwashing received frequently for the group leaders as well as their selected mentors. These leaders make it their duty to ensure that the relationship is properly maintained. The *pinnacleconductivity* is given to these complex effects which allow any of these individuals who by standard lie in the psycho-level zone, because they become highly *susceptible* to crime and violence as the conductivity of the charged ideas (*minutrons*) can easily be accelerated. These *minutrons accelerations* are constantly witnessed in group meetings as well as an active experience of reading propaganda books or watching films regarding preaching being delivered by master mentors. But, not all individuals can be encouraged to do something bad or evil; it all depends on the underlying character and the personality type of the individual. Some are more easily motivated than others as well as some are bolder than others. Some are more educated than others. The educated person may not be easily motivated to commit crime or engage in violence. Those that have been habituated to take drug substances during meetings may be more susceptible to violence than others.

The most spectacular property of the ***pinnacleconductivity state*** of superiority complex, and the one which provides it name, is the fact that resistance to violence or crime appear to be exactly zero. While the probability of irreversibility is higher ($p = 1$), resistance on the other hand appears to be zero ($p = 0$). This is a unique contrast between the ***superfluity state***, where the probability of the normal individual's resistance to crime or violence is, of course, higher ($p \geq 1$). This is because the mental ideas of the latter are not highly charged with false ideologies that can easily be accelerated by preaching or spontaneous influence or brainwashing. Thus it seems that with the superfluity state, the individual's mental ideas are ordered and somehow insulated against any external mechanism.

The Six Important Formal Axioms or Principles of Superiority Complex

Our procedure here will be to summarise the general principles of superiority complex theory in a simple way by using the language that linearity and non-linearity has provided us. The statement of each principle will be followed by some comments, for the sake of explicitness and emphasis.

Axiom 1

Associated with every psycho-superiority complex individual is a master mentor, such that the former's visualisation of the latter or constant attachment to him reinforces this behaviour that can cause the abnormal mental states. The mentor can be an individual or an organisation.

The cause-effect structure provides us with information where the complex individual originally contracts his false mentally-charged ideas (*minutrons*). The mention of a mentor can here be taken as a scholar who has published some books about these ideas, a well-known politician who has formulated a theory regarding this, or a very strong organisation that has been established to propagate these false ideas. This master mentor can be worshipped or venerated, and so the individual becomes engrossed in something that has a deep root and will not be easy to uproot. This will be worse if he got infected during the formative years of his life.

Axiom 2

There is association between the superiority complex and the kind of continent the individual hails from. While not making a specific statement regarding this, it can be asserted that race may have relationship with superiority complex.

This second principle provides a clearer interpretation of the superiority complex theory, in that it predicts that superiority complex has association with race. Furthermore, this principle predicts that superiority complex has affinity with continent of origin. These details of the relationship of complex are not needed for our purposes. They are to be preserved for future scholars who will eventually study the implication concerning the falsity or truth of these predictions.

Axiom 3

If S is an obsessed individual with superiority complex, the probability that D, a son to S, will turn to be a superiority complex individual will be high. D's actualisation of this condition will later be that determinant to his becoming a member of an underground organisation.

During their formative years children are influenced a great deal by their parents and significant others, such that in most cases the parents' professions can influence the offspring to go into the same profession. Complex ideas that have been transferred to children through indoctrination have caused some young men to commit murder in the racist south of the USA. There is a higher probability that a child grown with a racist parent will cultivate hate for others. A strong prediction can be made in connection with this law because there are large bodies of empirical materials that give support to this.

Axiom 4

Where superiority complex is the norm of the society, the probability is that the number of individuals suffering from it will be greater, and there will not be any difference between the normal perception of self from the abnormal. In other words, a distinct demarcation between normal and abnormal cannot be easily drawn.

Certainly, a higher proportion of individuals living in a society where these ideologies are regarded as a way of life will be found. This knowledge we obtain from group psychology and the theory of modelling advocated by Albert Bandura in his social cognitive theory. Through the years, modelling has been acknowledged to be one of the most powerful means of transmitting values, attitudes, and patterns of thought and behaviour. Following conceptual traditions, many theorists have conceptualised modelling as imitation. This is the process by which one organism matches the actions of another, usually close in time. Drawing on the psychodynamic tradition, a number of personality theorists and developmentalists interpret modelling process as identification. Identification is the adoption of either diverse patterns of behaviour, symbolic representation of the model, or motives, values, ideals, and conscience.

Axiom 5

Whenever a segment of the ruling class/figure becomes obsessed with superiority, the inhabitants suffer and this is particularly seen with foreign inhabitants that sojourn in their midst, and the condition of the ruling class/figure if not meticulous checked or controlled could lead to blunders or a catastrophic end by that community to its surroundings.

Numerous cases or facts can be drawn from history: Case 1. King Nebuchadnezzar, the Babylonian King and his constant brutality towards his own people and the Jews that was in captivity. He wanted to become like God, the most High and so he built a huge statue to represent himself. This king was obsessed with superiority. Case 2. Some Roman Emperors who became deified and wanted to remain like gods. The sufferings of the citizens of Rome and the Christians that lived in this Great Empire. Case 3. Some Persian Kings during the Great Persian Empire. Case 4. Alexander's obsession with superiority and his own understanding that he was a Grecian god incarnated. Case 5. The Second World War and the sufferings of the German people and the Jews who lived among them. Case 6. Modern Iraq under the rule of Sadaam Hussein and the sufferings of the Kurdish people. There were numerous examples (i.e., an English King) where some kings were detected to be not in contact with reality and yet these same kings in their "omnipotent" condition were Commanders in Chiefs in great wars that left millions of innocent lives to perish.

Axiom 6

Those individuals posing as "hidden forces", which operate from several corners in the complex societies, and function in high and low areas in the society on causing distractions, sabotages, and disturbances, shall afterwards become the same ruling class of the society.

Certain people that form an important segment of the ruling class engineer a great deal of problems that are seen to be ensuing. However, as their operations are not open to the public the authorities or the police cannot do anything to stop them. These activities are not different from the work of the secret police, though in the case of the secret police they have official permission from the state to do whatever they set out to do with the purpose of protecting the state.

Thus the **hidden variables** that provide different types of motivations to the individual with superiority complex are the *mentor-hidden variable* that comprises an organisation or a significant other. This variable seems to possess the central framework of the theory of superiority complex, and it is asserted to be the "first cause" that impressed the acceleration of this dynamic philosophy. The rest of the hidden variables are *the parents' hidden variable, the gang hidden variable,* and *the racist books and propaganda advertisements hidden variable.* The expression "hidden variables" refers to the manner these variables that are present to the individual seem to operate on this individual without him/her realising its destructive power. Remember that we have already asserted that many of these ideas are however, not known by the individual as false. It also points to the mechanism of entangling the individual, thereby allowing his/her rational faculty to become helpless without adequate control over it.

Complex and the Principle of Self-sufficiency

When we observe the societies in which the *subjectus* has his domicile and yet its equilibrium is not balanced, we can rightly make the conclusion that the major reason why higher equilibrium has not been reached is due to the principle of *self-sufficiency reason*. This statistical law states that societies that think themselves adequate in both human resources and natural resources usually feel some kind of contentment that makes them not allow others (foreign organisms) to participate in their nation building. Their feeling superior naturally bans any individual that is not of their kind to contribute in its nation building. Why they cherish that kind of feeling, though, may be justified at present time; however, history casts a shadow that is rightly the opposite of their arrogant behaviour. Usually they have built their country not by themselves alone, they had in fact received enormous helping hands from abroad, and materials from other people from different corners of the earth. And in order to cover what they might have thought as shameful or for sheer reason of keeping its history as secret, they pretend that they do not need anyone to give a helping hand. But their development, like all other developments that occur in different parts of the globe, have received some impetus, directions and physical support otherwise it could not have materialised. Such arrogant attitude is in contrast to the societies that rather boast of their integrated work or co-operation with all other nations from around the world. The superiority complex that originates the self-sufficiency can sometimes have unprecedented bad effects upon any developed society, and in the long term can be detrimental to foreign policy and continuing development.

Complex and Recognition

To this we shall add the intent for *recognition*. The origin of self-sufficiency might have been sparked off due to the resolution to obtain world recognition. In other words, the yearning to gain attention from the world around, and recognition in the world of competitors for supremacy, might have caused some societies to pursue a policy that had been habituated such that they find it difficult to retreat. Recognition may imply that a certain *false perception* of having already set a unique standard has been cherished or has been maintained, and that an attempt to relinquish may create psychological conflict leading to inferiority complex. As many *subjectuses* cannot accommodate these principles because it induces conflicts and low self-esteem (self-esteem can stem from evaluations based on competence or on possession of attributes that have been culturally invested with positive or negative value; in esteem arising from competence, people derive their sense of self-pride from fulfilling their standards of merits), a tangible response will be the best alternative, that is, to eradicate the strange and perplexing situation in order to maintain appropriate equilibrium level in the society.

Complex and the Selfishness Theory

In the above discussion we mentioned the isolationism reason and a few other reasons such as poverty, not having been oppressed before, and others that make a group become more fearful toward intruders into their territory. It is understood that these factors originate the superiority complex in the first place. Another reason that I propose here as serving to augment the reason why a group or individuals can be said to contract the complex disposition is the selfishness principle. This can be seen as the underlying factor that makes any group that migrates to distant places away from the masses or others do it in the first place. ***Selfishness is defined as deficient in consideration for others. It is concerned chiefly with one's own personal profit or pleasure.*** It is also actuated by self-interest. In a motive it only appeals to self-interest.

The theory state[21] that people that migrate to and live distant away from the main group or the total population is usually motivated to do so because of other reasons, of which selfishness is one. They may do that because of the realisation that they have potentials that either make them sufficient to survive alone away from others, or they have found something that they would not like to share with others. It could also be that they prefer to be alone, away from the troubles that are usually caused by the great majority that lives in the hordes. Now, before any such bold attempt of migrating would be undertaken, the individuals might have weighed themselves about their own potentials and have found that they could indeed survive without the help of others. In short, in general people that make such attempt are always found to be selfish, adventurous, and able to take risk. Therefore, a general selfishness theory of complex contends that people that lived in isolation for a long time had the selfishness disposition ingrained in them many thousands of years, such that it requires an enormous amount of time for them to work upon it if they intend to relinquish this vice.

In this modern day the selfishness vice can be inversely revealed in the domain of politics as ***solidarity***. Therefore, societies that have strong solidarity are somehow perceived to be indescribably selfish in the manner they conduct their affairs in general. The general thinking that prevails in the society is that solidarity is needed for us to be able to work together as one people. By maintaining this approach, they close doors for the minority that had migrated to stay among them. The group solidarity function strongly in the manner they take their decisions. It is not uncommon to hear from people saying this about them, that they are the "culture of consensus," which means for them the majority decision is always cherished, sometimes at the expense of the minorities' interest and feelings. When decisions are usually taken using the consensus yardstick, there is the propensity to neglect the views of the minority that do not agree with what the majority says. It attributes a crucial role to conformity in this field, and thus to majority pressure. Yet we must remember the meaning that many institutions attribute to consensus. It is valid if it is obtained under appropriate conditions. In particular, sufficient diversity must be ensured so as to mitigate any open or covert pressure, and to guarantee freedom of expression to everybody, so as to obtain agreement *within* the group and not *by* the group. In short, selfishness theory can be argued as playing a bigger function in the area of consensus, and decision taken by the superiority complex group or individuals. I adopt the three propositions on which social psychologists have also agreed that they usually prevail in decision-taking by the consensus societies that is also found in connection with superiority complex groups or individuals:

Proposition 1

First, there exists a positive correlation between the positions originally adopted by members of the group and those on which this group reaches agreement.

Proposition 2

---

[21] Some will question that, are not countries that give aid or charity so kind and therefore in this case cannot be counted as selfish? My answer is that giving charity or aid do not make a country generous as these aids are given for political reasons and in most cases the money that are being given out are in anticipation that trade with a particular nation will continue to flourish. As a result aid or charity giving cannot be used to test the generosity of a nation.

The direction in which these positions are polarised is determined by the dominant values in the population to which the group belongs.

Proposition 3

Finally, during the discussion the group members become involved collectively and change the way in which they judge, perceive, and choose, depending on whether they are close to the dominant or subordinate pole in the scale values.

In the interim, I want to leave these propositions as they are without giving any further comments but instead ask the question, "What does consensus mean?" What distinctions can we make between closed and open societies and which of these two seems to be the breeding ground of the superiority complex individuals?

Consensus

Consensus can be defined as a general agreement of a testimony, opinion, etc. It simply refers to the majority view or collective opinion. It comes from the Latin word *consentire,* meaning "feel." Consensus is the decision of the majority that also affects the *subjectuses* in societies that have been known for their adherence to collective decision to run their affairs or politics. Where the collective decision or majority views is the norm for a better decision, it can be carried to the extreme in order to jeopardise the rights of the minority in which the *subjectus* is part. What kind of character does the society that frequently uses the norm of consensus made of?

Closed and Open Groups

Let us briefly discuss the distinction between closed and open societies and how the closeness a particular society has can be asserted as supporting the theory of superiority complex. The distinction, which was first published in a masterpiece, *The Two Sources of Morality and Religion*, in 1976, could be utilised to elucidate the kind of society that breeds the superiority complex individuals. This has been discussed well in the work of Moscovici & Doise. Thus according to Bergson, closed societies rally their subjects around obligations and special rules, conditioning a form of thinking and human behaviour that allowed no deviation. The open societies, on the contrary, originate because of the attraction subjects feel for one another and a capacity to elevate ideals to be pursued. The closed societies urge adherence to certain ideologies and a distinctive type of personality, leading certain subjects to cleave together, on condition that other individuals, branded to be alien or enemies, are expelled. The open societies elevate what Moscovici & Doise has called "an exemplar of universal validity," this could be a saint or the citizen, in whom each subject can visualise himself mirrored. Thus, in the latter case (open), a man is for another man a fellow countryman or compatriot, and in the former (closed), a fellow human being, that is, tribesman, or one of the human species.

    One comprehends from Bergson's analysis that closed societies can consist of families, clans, churches, and also nations. Irrespective of their size, these organisations cling to the same authoritarian hierarchy that makes them possess a similar mental stereotype, as well as the same religious faith. Above all, these organisations tend towards the same ethnocentricity, for they consider the values of their society as superior to all others or rather unique. This makes them point their contempt or hostility towards certain neighbours, groups, families, clans, and others in their vicinity that differ from them. Even in this practical world some tribesmen regard their members alone to be the "right," "good," "excellent," people. It is only their members who have the endorsement to be regarded as real "men." All members of other clans are to be excluded from the human race, for they are "monkeys," "pigs," "jungle dwellers," and etc. As praise and honour are showered upon themselves, on the other hand, hate, insults, and abhorrence are directed against others. These two hyperboles' relationship, according to Moscovici & Doise, is condensed into a statement issued by Bergson as this: "Let us merely say that the two contrary maxims, *homo homini deus* and *homo homini lupus,* are easily reconciled." By the mention of *homo homini deus*, a subject may be referring to his fellow countryman or his compatriot, while in the case of *homo homini lupus,* his mind will be focusing on foreigners or aliens.[22]

As has been pointed out, closed human groupings are made up of the family, the nation, or the party. They form their own reference groups, and their members consist of those who identify with them, and they resemble one another. The relationships with closed groups differ from those in the open groupings. Whereas in the closed grouping their relationships are made of exclusion and discrimination, the open group ones are characterised with contract and recognition. The closed group possesses the superior idea that usually makes them see themselves as the "chosen people." It is this same notion that is transposed to nations or families, allowing them to perceive the "social identity" or the good bond that prevails among its members. This notion can be carried further to the extent that members of these groups that perceive themselves to be superior can on this very basis declare that same superiority in a subjective manner; i.e., for themselves individually. On the contrary, open groupings revere the balance of interests and rights which emerge as a result of the combined power of each individual, "the compatibility between how they represent these interests and rights, and upon society in general."[23] This is all the more reason why in the context of this objective world, closed groupings are considered as living in different worlds, while the open groupings live in the one, same world.

It seems that Bergson's work supports us regarding the breeding ground of the superiority complex individuals as situating in the closed society more than that of the open society. Their manners of worshipping consensus decision-making can create problems for the minority that inhabit within a group or society. However, we do not generalise in this particular instance, as there can be some open societies that may deliberately champion such an approach for political reasons. The fact that some closed societies do this does not mean that all people living in that society cherish to be viewed as superiority complex, and therefore adhere to the principle of using the consensus yardstick to make an important decision.

Scholium

The theory of superiority complex has been developed purposely to portray the salient behaviour of certain individuals whose obsession with themselves as superior or better than others compel them to cause injury or disturbances to other people elsewhere in the world.

Reasons as to why individuals do not reach their goals in a state where the equilibrium is low have also been described above in this theory. With this we get the impression that the *subjectus* has not been well assimilated into the society, and that he may be in an inert state of his life that is still full of energy but appears redundant. The psycho-

---

[22] Moscovici & Doise (1994).
[23] *Ibid.*

superiority complex individuals who oppose this excellent attempt to bring in the *subjectuses* may engineer certain manners of sabotages that may convey the confusing situation occurring.

What is unique about the inhabitants of these societies that possess superiority complex is that the strong solidarity they have make them not react even when they see that someone is maltreating a *subjectus* or blocking this person's chances in the society in question. In fact, they become exhilarated and always maintain a hypocritical smile as if nothing wrong was going on. The strength of this maltreatment is seen in the fact that when one cannot actualise his potentialities, one is forced to leave the country in question. It has become a successful way of acting to frustrate the plans of the ambitious individuals in these societies. The tension that results between many *subjectuses* and the superiority complex individuals are described to be something else rather than their own engineering vices. In other words, a strategy of branding all *subjectuses* as criminals, drug addicts, or sellers of drugs quickly project the real problem of the superiority complex individuals on to something else, and in turn the former will be branded as possessing insurmountable problems. This is now a classical view, as now the theory of superiority complex points out that the problem may well be from the superiority complex individuals who need some psychological treatments or conditioning in order to change a character so well ingrained in the persons found to be always innocent.

The strength of this theory lies in the fact that it unveils a major problem lurking in the dark, and this also sounds a warning with regard to the consequences of behaviour accepted by so many people as normal, though it may have some defects. When many people that we do suspect to be ill are now being diagnosed as unhealthy, it sounds a greater warning to us all. Failure to recognise this may continue to frustrate the lives of others, which may not enable them to function well in the society in order to actualise their potentials. The principle of non-discrimination law has suggested different approaches that will lessen the predicaments, which will make the *subjectus* his potentialities being actualised. But before we leave this theory, let us consider some important theorems that are the direct deductions from the axioms or the six principles postulated for the theory of superiority complex. For the great value of any theory lies in its power to suggest new laws that can be confirmed by empirical means.

Superiority Complex can be transmitted

The Conspiracy Theorem

Minutron theory

Definition: By *complex transmission,* we mean that any act that is unlawfully pursued or carried out by the self-exalted human being to injure another person or his life so that she/he will not be capable to function appropriately or enjoy his state of success or peace. In connection with this theorem, I also propose the hypothesis that states that charged complex ideas that has been termed "*minutron*"[24] can be transmitted to another person (this follows from principles 3 & 4). In other words, individuals can be infected with these non-observable abnormal mental states. In the context of the theory of superiority complex, let this complex

---

[24] Like electrons that carry electricity in solids, *minutrons* carry the charged ideas in the brain field as well as transmits charges through wave medium to another individual.

reaction field be formally known as the ***conspiracy principle or theorem***. And also let the process whereby the translation of ***minutrons*** between the inert condition and the kinetic condition be known as the ***bilateral intercourse***. Imagine a man Φ with considerable influence and power that occupies a high position in authority. Let ϑ represents his good friend, who in the world of influence and contacts has been aided to acquire a high position in the public sector. If the person Φ should one day have problems (that lead to loggerheads) with a third person Ω, who happens to work at the department where the person ϑ has influence or control and is acquainted with him. The man Φ will use his influence on the person ϑ to do unfavourable things to Ω that can lead the last-mentioned being sacked or removed from his wonderful position. We conclude at this instance that the psycho-superiority complex of the man Φ has been transmitted to the man ϑ, or the last-mentioned person has been ***infected*** with the first-mentioned person's madness. The momenta of these ***minutrons*** of the person Φ can be spread to different individuals, supposing the vice that has to be carried out requires the involvement of many people or depends on several techniques to be utilised. Here it can be further *hypothesised that the extent of the disturbance to Ω will be related to (i.e., proportional to) the moving energy of the **minutrons** to be transmitted from the person **Φ** to other persons involved in this conspiracy.* Where the pressure put on a person's ***minutrons*** is constant, the damage caused by the person Φ to the individual Ω will be irreparable. Momentum (G), which is like a punch and in physics is mass (m) multiplied by velocity, in our theory is the moving and exchange of contacts of ***minutrons*** between the psycho-superiority complex individual and other individuals. It is known in our theory as the *cross-contact fields of momentum,* $(G = \iota/\lambda)$ and their increments per unit of time are equated to the components of the power (**¥**$_E = \iota mc^2/\lambda^2$).

Some characteristics of the ***minutrons*** are that:

(1) they possess *mass* and *speed*;
(2) the acceleration of *minutrons* are unobservable;
(3) the reception of a *minutron* causes a reaction in the brain field;
(4) there is intensity of the charged *minutrons*; and finally,
(5) *minutrons* can be measured and require energy, in the form of ***mc**^{2}* and they can be comparable to electrical current.

The collision of *minutrons* occurs where the recipient of a plot rejects the plan orchestrated or suggested by a psycho-superiority complex person, which is seldom the case in communities where superiority complex ideas are the norm. Most collisions do change the force or energy in the *minutrons*. It has become necessary to introduce the term *minutrons perforation* to denote this state where collision occurs between a psycho-superiority complex person and a co-aid complex. But, as the situation predicts, there is some permanency of the easy flow of *minutrons* among the society that the complex ideas are the norm and this condition makes us to regard the *minutrons permeability* ($p = 1$) as a property of this latter stage, where suggestions of conspiracy found easy supporters among these groups of people. This easily suggests a kind of biased attitude inherent in this culture or society of people being discussed. The opposite of these circumstances is that where there seems to be uneasy access to *minutrons* of complex free society or semi-complex societies is described to be *minutrons impermeability* ($p = 0$ and $p < 1$ respectively).

Symbolically, the formula for this can be written like this:

¥ $= (\Phi\iota + \vartheta u) \otimes \partial > \Omega\iota$

Where ¥ represents a complex reaction field and ι represents the *minutrons* of individuals. Here Φι is the parent complex,[25] ϑu the co-aid complex, ∂ represents the work to be done (the work function of the *minutrons* or, in other words, the energy needed to cause non-equilibrium to the resultant complex), and Ωι the resultant complex, which is otherwise known as the sufferer's madness. Thus, when there is **(Φι + ϑu), ¥ = 0**, which indicates no reaction in the complex fields. But **(Φι + ϑu) ⊗ ∂** means *minutrons* are charged, for there is some work being done. This can be represented by θ. Where we therefore have **(Φι + ϑu) ⊗ ∂ >Ωι**, it indicates that **¥ = 1**, a strong indication that complex reaction is going on. It also provides us the complete complex reaction fields. The co-aid complex can be increased to encompass many different persons, depending on the magnitude of the problem and the work to be done to enable the plot to be accomplished successfully [i.e., **(ϑu1+ϑι2+ ϑι3+ϑι4...ϑιn) ⊗∂**].[26] Between the superiority complex, that is, the parent complex, and a co-aid complex/complexes, the *minutrons* travel in wave mediums. This complex relation holds, in the whole world of human relations, for where there is no conspiracy the complex action field reduces to **(Φι + ϑu)**, which is **¥ = 0**, a normal return to the amicable relationship.

To employ the calculus of classes, we pursue this formula:

¥= (Φι + ϑu) ⊗ ∂ >Ωι
¥= Φ∂ι + ϑu∂ > Ωι *by principle of distribution*
¥= Φ∂ι + ϑu∂ > Ωι *by principle of simplification*
¥= Φ∂ι + ϑu∂ > Ωι

Where here the relation can be expressed as Ωι is "included in" **Φ∂ι + ϑu∂** or **Φ∂ι + ϑu∂** is "greater than" Ωι. It can still be expressed as Ωι is "equal to" **Φ∂ι + ϑu∂**. This theorem has wider implications, for example, in the world of wars and minor conflicts, where we can move from the personal experiences level to the societal level. There is no question about the fact that according to our theorem, **war is a symptom of disorder**. Since in the last hundred years the two World Wars we have fought have unveiled how the complex madness of one man or men had led the whole world into going to war. A classic example is seen with a cherished and famous war leader who lived in the middle of the twentieth century. Unfortunately, his ruling years were brought to an abrupt end through assassination. It was told by some of his aides that when the drugs that were supposed to cure his troubled illness were injected he felt like an "omnipotent." This made many of his decisions that he took inappropriate, and led to a catastrophic end of the war he led his country to wage. This is but one of sundry examples that many countries of war had witnessed. While these leaders managed to infect some people with their madness as accomplices, they also succeeded in transmitting their illness to the sufferers or victims, whose tortures till death preceded their madness of defending or causing untold havoc on the planet earth. Some madness can become permanent, especially the sufferers' madness. The reason why this conspiracy theorem is also important to the theory of superiority complex is that it aids us to comprehend what the objectives of those people entangled in its philosophies are apt to accomplish, and what those affected by it experience in their day to day lives. Sabotages and conspiracies are mirrored in the first-mentioned delusive states, as these help them in the augmentation and holding on to their illusory superior power. This theorem enables us to derive a new theoretical law to be known as the **Complex Action Law** which states that: *At any complex action fields there exists always the parent complex, which is the psycho-superiority complex individuals or individual, co-aid complexes or complex, that is, the accomplice and the resultant complex,*

---

[25] Many of these people are deluded into believing that they are superior or have been called to fulfil a promise.
[26] This follows addition theorems for distributions where these are seen as independent random variables.

otherwise known as the sufferer complex. The product of the work done by the parent complex-co-aid complex being greater than the resultant complex.[27]

If we want to embellish our law to encompass the speed of light and the wave medium which is common in Mathematical Physics, our new Complex Action Law formula could be written like this:

$$¥ = (\Phi\iota + \vartheta u) \otimes \partial > \Omega\iota \times (mc^2/\lambda^2)$$

$$¥ = \frac{(\Phi\iota + \vartheta u) \otimes \partial \cdot mc^2}{\lambda^2} > \Omega\iota$$

Where c^2 is the speed of light, **m** is mass, and λ^2 is the wavelength or the distance travelled by the *minutrons*. This formula is valid and more appropriate if we understand the fact that *minutrons* only get to others through these faster mediums.

Still our formula can be reduced to appear like this:

$$¥_E = \frac{W\iota_n mc^2}{\lambda^2} > \Omega\iota$$

Where ***E*** is energy of the complex reaction, $W\iota$ is the work done; **n** is any integer from 0 to ∞, c^2 velocity of light and λ^2 being the wavelength, that is, distance travelled by the *minutrons*.
Q.E.D.

Minor Theorems

There are other theorems which are also of considerable importance in the theory of superiority complex. In the second theorem, here I emulate the same principle that made Albert Einstein, one of the greatest physicists in our time, to attribute mass to energy (i.e., energy possesses a certain inert mass). ***That not only has [the individual who possesses] complex ideas some disorder, but where there is disorder there must also be complex ideas***. Let complex ideas be represented by **C**χ and disorder be represented by **D**∂. Then the equivalent equation being discussed above is **C**χ = **D**∂. This second theorem is obvious and follows from principle 1, 5, 6 and Def. 10. This theorem justifies us in considering the complex ideologies of most war leaders that did not heed the voice of the international communities as entailing some kind of disorders. This interpretation agrees, moreover, with the picture of the complex madness revealed in wars that took place in the ancient kingdoms such as the Babylonian, Persian, Grecian, and Roman Empires. If we follow the theory of superiority complex, we might also hypothesise that the energy needed (the time spent for peaceful means and deliberations) for persuading parent complex and co-aid complexes in not going to war is the greater, the more disorder/complex ideas the parent complex fellow contains. The increase in disorders are not only due to the complex ideas imbibed from these mentor figures/background influences, but also the energy content of the norms mediated by the society.
Q.E.D.

---

[27] In one sense this is an empirical law gained through analysis of a theoretical law. For a theoretical law helps to explain empirical laws already formulated, and to permit the derivation of new empirical laws.

We can still add a third theorem: *Let normal individuals live long in communities where superiority complex ideas are the norm. They will tend to condone, and in this process come to identify with these ideas or conform.* To prove this, briefly we shall state that if χ represents the complex ideologies and μ the normal individual, in the course of time τ by living among the members of the community with χ ideologies (μ) will become $\chi\,(\mu)(\tau)$; that is, $(\mu,\tau)\,\chi = \chi\,\mu\tau$, or the complex oriented individual. This follows from principles 1, 3, and 4. The Second World War offers us a classic example. During the war many people that moved to central Europe who had previously spoken against the complex ideologies found themselves becoming adherents (principle 6). Some were later sent as ambassadors to different parts of the planet in order to propagate these complex ideologies to other new converts. Those who did not identify with these complex ideologies later conformed and even helped in betraying innocent people into death camps.
Q.E.D.

The fourth theorem states that: *Superiority complex ideas exist in all human beings universally, and their intensity is proportional to the increase gained in knowledge.* This follows from principles 1, 2, 5, 6, Def. 10, recognition principle, and also the law of the principle of self-sufficiency. We have already proved and further pointed out how this is possible for not only people from continents that have high advancement in knowledge and technology who tend to cherish complex superiority ideas over others, but also individuals from less developed nations also do that. It follows, therefore, that apart from this generalisation of certain peoples that have high technology and regard themselves as distinct, other individuals, irrespective of their continent of origin, tend to cherish complex ideas when they have acquired extensive knowledge that puts demarcation between them and others who possess mediocre education. This becomes like a master-servant relationship, boss-subordinate relationship, and superintendent-vassal state relationship, as are commonly observed in societies where they practise the cast system
Q.E.D.

The fifth theorem: *If two individuals, **P** and **J**, are constantly attached to a master-mentor and a significant other respectively, I say that they will be influenced equally in proportion to time and inversely proportional to the quantity of ideologies.* For if the individual **P** has made in his visualisation the master-mentor **A** as the primary object that influenced his life, it is likely that he has a significant other **B** who may be his father, a member of the family, or a member of a gang, that has been established purposely to propagate these complex ideas. By the additive rule **P** has influencing factors **A + B**, which comprise the master-mentor **A** and a latent significant other, **B**. Let **B** be the significant other for **J**, who is the major influencing factor for **J**. Since by principle 1 any superiority complex individual always possesses a master-mentor that is visualised and venerated by him/her together with the group, **J** may certainly have **A** or a prototype as his master-mentor. Individual **J** may in the end have **B + A** as his influencing factors. Therefore, equal influencing factors for **P** and **J** will be:

A + B = B + A

Thus, influence (*I*) for P is $I = (A + B)$, and for J is $I = (B+A)$

But since according to principle 1 there are different master-mentors, for example, a scholar-mentor, a famous politician-mentor, or an organisation-mentor, and etc, ideologies may differ. This means in terms of complex ideologies, increase with time will not necessarily be noticed with these individuals (i.e., quantitative or magnitude of ideologies). Moreover, the different quantities of ideologies that exist in the complex fields may not be present equally to **P** and **J**;

one may obtain more ideologies than the other, as well as faced with radical notions that are not usual with all master-mentor or mentor organisations. We can conclude at this instance with certainty that the quantity of ideologies may be inversely proportional to time with regards to complex development of **P** and **J**. That is, the former may decrease, though that will not affect the progress or development of the latter. This theorem follows from principles 1 and 3. To employ the second order differential equation formula, our theorem may be written like this:

$$\frac{d^2p}{dt^2}(t) = I(A+B)(p)(t) = -\frac{d^2\varpi}{dt^2} = I(A+B)(p)(t)$$

$$\frac{d^2j}{dt^2}(t) = I(B+A)(j)(t) = -\frac{d^2\varpi}{dt^2} = I(B+A)(j)(t)$$

Where I is the influence on the individuals at the time t and ϖ is the decrease in the rate of growth of ideologies. Time here is said to be causal, at the state of t_o, it is possible to predict exactly what state will be at any later time t.
Q.E.D.

Corollary 1. If the complex ideologies do not grow or increase with time as the intensity of the influence, then these ideologies may consist of childish ideas and infantile notions that may not easily offer themselves to critical thinking of the mature men.

Corollary 2. Hence immature ideas and infantile notions consist the complex individual's thoughts, which the critical-minded person will not entertain.

Finally, the sixth theorem states that: *With the same conditions being supposed as in theorems 1, 2, 3,4, 5, and Def. 10, rather than being a sign of greatness war is a symptom of a disorder.* Let all that have been asserted be summed up together, there are convincing arguments and a large body of empirical evidence that support this theorem that war is a disorder. It has always been thought and confirmed even way back to the Roman Empire which lasted for over two thousand years, the longest empire so far since the inception of the world. The brutalities of these emperors or leaders were always associated with madness, or some kind of illness, and they did not end up there, they saw to it that their soldiers commenced unnecessary wars when they were in these unpredictable mood. It happened during the time of the British Empire as well as during the German aggression. History has recorded numerous wars all over the planets where leader madness had preceded the inception of great wars.
Q.E.D.

Corollary 1. Everything that has been demonstrated above about the transmission of complex or infection of complex to healthy innocent individuals around the world holds when the psycho-complex individual passes on his complex to co-aid complexes, and later to all the sufferer complexes.

Corollary 2. Therefore some important figures when contracted their obsessions with superiority find pleasure in causing injuries to innocent healthy people.

Corollary 3. So far as it is the disorder that originates the instability among people and transmits disturbances to the greater majority, obsession with superiority cannot be considered as a sign of greatness but a disorder indeed.

Minutrons Perforation

Our next object is *minutrons perforation*, as appearing in the theory of superiority complex. We have defined this already as the state where collision occurs between the *minutrons* of a psycho-superiority complex person and a co-aid complex. This make a plot lose its force or energy. The speeds of the *minutrons* are impeded, and usually such occurrences break away any conspiracy in the world of influence or war. How does this occur, and does it truly happen among individuals in the complex societies? What is the result of this split, and does this contribute to normal relationship between people that are in friendship relationship or alliance agreement?

Minutrons are rejected, collided, and *impermeability* becomes high and can be equal to one (i.e., $p=1$). This signifies a sign of no interest of any conspiracy whatsoever. Yet with some co-aid complexes it may signify *minutrons displacement;* this can denote a state of indecision or redundancy. To make the *minutrons* gain the lost interest or energy, the pressure had to be increased, as well as reinforced with rewards. At this moment, there is mistrust on the side of the psycho-superiority individual regarding the co-aid complex/complexes. To insure that trust is returned or regained a kind of compromise had to be reached. In the case of war, *minutron perforation* would not be necessarily an end, nations would be compelled to support in a passive manner by sending support in the form of weapons, or fiscal cash and so forth. This passive succour of a plot can be regarded as complex disposition, and can be comparable to the ***"double standard game"*** that we see in wartime by the so-called countries that call themselves non-alliance partners or neutral nations.

In the personal experience level people with high education, especially university professors, scientific geniuses, and gifted politicians can be deceived to succumb to the clever manipulations orchestrated by the psycho-superiority complex individual. The university professor or scientific genius would utilise his intellectual resources to perform indecent experiments for the psycho-superiority complex individual to help win war or succeed in a conspiracy. Even a gifted politician would sacrifice her several years of experience and wisdom acquired and then turns away temporarily against an ideology that she/he had believed in for years. If it were equality of men, she would condone with the interpretation offered by the psycho-complex individual. Others would join the sick psycho-complex individual in their intimidation, sabotaging, appear to be siding just to assuage the disgrace that a psycho-complex individual they support is going through. The case of W. Heisenberg is a typical example; he sacrificed his giant intellectual mind to support a famous war leader of the Second World War. In the end this leader sacked a famous scientist who was his colleague at his department, but for fear of loosing his position he sided with this leader, or did nothing to prevent his colleague from being sacked. The psycho-complex individual in his delusion usually employs certain words such as "solidarity," "let's keep the tradition," "we are the best," "superior," or call certain names to degrade the sufferer; all are to justify the psycho-complex individual's position. The individuals that easily succumb to the influences of the psycho-superiority complex individuals are termed by our theory as *roboantropos*. Originally from the Greek language, *antropos* means "man" and robot implies that this fellow cannot think independently, he is like a robot that can be programmed by any psycho-superiority complex individual. Those who defiantly succeed in rejecting or evading the manipulations of a psycho-complex individual and not only that but also help others to unveil these kinds of manipulation which in the end save the lives of innocent citizens, are known as the *sokraantropos*. This is taken from the name of "Socrates."

Among those that are called *roboantropos* individuals, are a few extraordinary individuals who become totally dependent on the *minutrons* of the psycho-complex person, such that these persons (i.e., the former) whole life existence become attached to their master-mentor (e.g., Saul of Tarsus, the Inquisitor General who championed the Spanish Inquisition (Tomás de Torquemada), Heinrich Himmler and a few other aides of Herr Adolf Hitler, and the Witch hunters in the Middle Ages Europe, etc.). They were the *manipulation susceptibility minutrons,* those who depended on their supreme master-mentor for the "air they breathed," and could not do anything without their mentors' permission. Usually such people become the tools for destruction that take place whenever a psycho-complex individual becomes a leader, and through tyrannical means or war cause successive destruction to the physical world or to mankind. They normally follow the direct instructions of their master to the letter and never waver in their support and dependence upon him. One of these individuals asserted that he could hear some voices telling him always in his own ears repeatedly that "the Jews were not human beings." The delusion people of this kind suffer is unprecedented and cannot be described in words. In both instances, the *manipulation susceptibility* Saul of Tarsus could persecute the Christians and later could as well become manipulated to carry out again the only message he could hear from his master-mentor, the resurrected Jesus Christ, when he was on his way to Damascus to obtain permission to enable him persecute the Christians all the more.

Let us try to put into a general mathematical formula what actually happens when *minutron* perforation occurs in any neighbourhood. For it is possible, however, to extend this disturbance to a more general problem in which the splitting potential undergoes large changes, but over such a long period of time that the change in potential during the period of the *minutrons* acceleration that is channelled in transition to the neighbouring state is compared with change of energy involved in this passing from one to another. To observes what happens precisely, imagine there are:

$$\frac{\Upsilon \, \partial mc^2}{(E_s{}^\circ - E_n{}^\circ)\partial t} \ll 1$$

where the suffix s denotes the direction along the movement of *minutrons,* $E_s{}^\circ$ is the initial *minutrons* energy from a psycho s complex, E_n the energy of the co-aid complex in any neighbourhood or neighbouring state, and Υ the period of the transfer of *minutrons* in question.[28] Supposing the period of the *minutrons* accelerated in transition is $\Upsilon = \iota / (E_s{}^\circ - E_n)$, our complete formula of any *minutrons* rejection, disturbance or a split will be like

$$\frac{\iota \, \partial mc^2}{(E_s{}^\circ \neq E_n{}^\circ)\, \partial t} \ll 1$$

This, in fact, means that $\partial \mathbf{mc^2}/\partial t$ will be equivalent to zero ($\ll 1$) as the $\neq$ in the formula also indicate no *minutrons* reception on behalf of the co-aid complex.

Scholium

The six axioms and the six theorems which we have demonstrated rigorously by their aid enable us to enumerate all the important laws of the theory of superiority complex. As the reader will notice, the theorems are necessary consequences of the principles, so that if the principles are accepted, the theorems must be accepted also. I believe I have managed to provide sound explanations to some of the important questions which this theory has raised, and it is my earnest desire that the laws generated will be of good use in this modern world.

---

[28] Like electrons, velocities of *minutrons* in atoms are much higher ($\cong 10_8$ cm/sec).

And that, without prejudice, all people who happen to read this theory will lend their ears and also examine the theory critically to see how its comprehension will enable them to desist from certain behaviours that cause injuries to their neighbours. The theory's great predictive power should allow us to accept it, as well as test it, for its truth or falsity, and consider it as one of the most useful theories in the twenty-first century.

Miscellaneous

Here, I provide other important theorems, which will be proved or demonstrated at a later date.

Theorem 7

Psycho-superiority complex individuals can impede the progress of certain individuals or nations.

Theorem 8

Psycho-superiority complex individuals' disorders cannot be cured because they are not aware that they have the problem.

Theorem 9

The underground groups who suffer some form of psycho-superiority complex disorders are made up of a segment of the ruling class.

Theorem 10

All leaders with superiority complex problems or disorders delight in causing disturbances among their own subjects and other inhabitants outside their dominion to prove their powers.

Theorem 11

All psycho-superiority complex individuals cause disturbances to those who pose a threat to them and that without this threat no psycho-superiority complex will engage in crime or disturbances.

Theorem 12

Those major hostilities that lead to the hampering of progress caused by the psycho s complex individuals to other individuals or groups usually leaves its deadly mark and wounds on the planet earth.

Theorem 13

The causation of disturbance by psycho s complex leader to his subjects or citizens, and its consequent transference of these hostilities outside the leader's dominion are the sources of aggressions that lead to catastrophic World Wars on the planet earth.

Theorem 14

If a segment of every ruling class suffer some form of psycho s complex disorders or disturbance, then this may well explain the brutalities of many modern governments as well as ancient governments brutalities to its subjects or other subjects outside their dominions.

Theorem 15

If the complex ideology of psycho s complex individuals does not grow in relation to their mental faculty, then this may cause conflict with their mature minds that may lead to the possession of insecure behaviour.

Theorem 16

The transmission of superiority complex disturbances/disorders to other innocent healthy individuals or groups in this world is the source of chaos on this planet.

Theorems 17

If individuals who possess the superiority complex ideas have disorders, then they may commit crimes without knowing.

Theorems 18

If those normal individuals who come into contact with complex ideologies/societies condone and through this process conform or identify, then they may indirectly be supporters of those militants groups in these societies.

Theorems 19

The psycho superiority complex personalities and their ideologies, which are found in all societies, may be the carriers or sources of all common conflicts that occur around the globe.

Theorem 20

The psycho s complex individuals' disturbances to those who pose a threat to them make the world not a peaceful place to live in.

Theorem 21

If individuals who possess the superiority complex ideas have disorders without knowing, then they may be susceptible to criminal tendencies that may cause instability.

Theorem 22

World wars are always caused/originated by insecure personalities who have superiority complex disorders, and since they are usually not detected, are transferred to innocent individuals through the help of co-aid complex personalities, and these wars cause havoc to our physical environment and the planet earth at large.

Theorem 23

To the extent that war is originated by psycho s complex personality-type, war can be profoundly described as a symptom of a disorder, though to the sick psycho s complex person war is seen as a sign of greatness.

Theorem 24

It is the society as whole that moulds the psycho s complex personality-type just as it helps the forming or moulding of the criminals among us.

Theorem 25

If the sick psycho s complex inappropriate behaviour originates war, then war can be regarded as a symptom of a disorder.

Theorem 26

Master mentors are always venerated or worshipped and their constant visualisation or attachment causes the abnormal mental states in their followers/dependants; and these dependants do not usually perceive these miserable conditions as so.

Theorem 27

Of all master mentors, those that are not perceived through the sense organs are the most common ones, and their grips on those that depend on them are most profound.

Theorem 28

Human mentors who are dead usually had left their legacies, which are in the forms of books, ideologies, and many memorable things such as their burial places or shrines.

Theorem 29

If master mentors are neither human agents nor organisations, but instead the unseen agents, then their influences are bound to be everlasting ones.

Theorem 30

If the complex ideologies are the norm of any society, then there will be certainly more people in the category of abnormal than the normal in this particular society.

Theorem 31

If complex personalities or ideologies are mediated through parents and other significant others to the individual, they will have master mentors always as additional reinforcers in these societies.

Theorem 32

The underground psycho s complex individuals form a segment of the ruling class, and their functions, with regards to complex transmission, are seen to be a criminal offence.

Theorem 33

The transmission of superiority complex disorders to innocent healthy individuals is a criminal act.

Theorem 34

To pose a threat to someone does not give a license for being attacked by the latter individual; but if an individual is attacked or harassed because he is source of a threat, then the doer of this deed is a psycho s complex individual and his act is regarded as a crime.

Theorem 35

A psycho s complex is not born with his illness; in other words, obsession with superiority is not a hereditary illness though it can be infected or transmitted.

Theorem 36

Because of his illness condition, a psycho s complex individual can commit a crime without being aware of it, and he could reveal inappropriate behaviour in public without knowing as well as boast out his secret plans to his enemies without being aware.

Theorem 37

If the psycho s complex obsession concerns race superiority, then it will be impossible to recognise all people on an equal basis, as he will suffer from an even greater disorder of discriminating others he considers as inferior.

Theorem 38

It is the intent to dominate all others that makes many psycho s complex individuals become obsessed with race superiority, and since race superiority is false, they must result to criminal tendencies, such as war, aggression, and other propaganda tactics to accomplish their aims.

Theorem 39

Of all other reasons that generate the obsession with race superiority, it is insecurity with selves that originate the obsession with race superiority, and since this is a form of illness can easily be transmitted or infected.

Theorem 40

If the contents of the ideologies race superiority persons (obsessions) hold consist of infantile thoughts, then they may err without being aware, and this will be compounded with insecure behaviours in their own milieu.

Theorem 41

Obsession with race superiority is the result of insecurity with selves due to the threat the psycho s complex individuals face from without, and this may compel any race to invade its neighbour's territory, which is a criminal act and evidence of madness.

Chapter 7: Discrimination and the Theory of Alliances

The Theory of Alliance

Where the *subjectus* has not succeeded in getting himself all the things he wishes to acquire from the system or the society that he lives in, in some cases the alliance principle will provide a way out. Or at least it can expose him to important contacts in the country of his domicile. Most important of all, in case of any major **conflicts**, the alliance partners are the best organs to fight for a member and also look for a better settlement that usually suits both parties. It is a generally accepted fact that forming of an alliance with the purpose of defending one's nation entails some problems, but not in the area of getting co-ordination with other organisations with the purpose of acquiring material benefits and protection. The theory of alliance is usually discussed in the social sciences in the area of war and conflicts. It is a situation where independent countries or incoming states pair with each other to form a partnership for defence in times of war. In alliance formation, many countries look for stronger partners who have all the human resources and natural resources that can aid them to win war or defend themselves in case there is one. And so we often hear being mentioned of today, USA and its allies, Russia and its allies, the Communist bloc and its allies, etc. Usually the word "alliance" is not used in any technical sense, and therefore can be understood as less legal, it can mean any sort of amicable co-ordination of policy. Thus we find strong alliances like that between the USA and Great Britain, and weaker friendships alliance between, for instance, Britain and Ghana, where there are fewer formal agreements. Evidently, the borderline between an alliance and an agreement (that is, in the case War alliances) is a slim and wavering one. We can almost render an alliance as existing when there is some sufficiently large number of specific or implicit agreements.

   Let me say that the design of this chapter is to premise the work by way of describing and offering implicit definitions, and later present the axioms. These will then be proved by reason and observation, and if necessary with experiments conducted by others.

The Necessity of Alliances

Alliance can be the only way many foreign organisms can get acclimatised in the system, and therefore it is imperative if the *subjectuses* at early stages in the society in question do more research about it. The principles enumerated in the relationship of war partnership also hold well with the alliances in the different societies. Some alliances can be friendship motivated and therefore can be regarded as friendly alliance to distinguish it from the more formal alliance, where the agreements may be many, and also ensures deep commitments. In some societies, though, the authorities do not encourage the joining of certain alliances, yet the degree of willingness of its natives themselves to be part of an alliance signal a warning for the *subjectuses* to do the same. The benefits are enormous but the costs can be very high in terms of money; that is, fees for the alliance and time to attend all its numerous meetings. The purpose of the theory of alliance is that it enables the *subjectus* to become part of a group, organisations or a partnership that he can count on. These can aid the *subjectus*, say, in looking a job, fighting for promotion, fighting for discrimination in jobs, looking for financial, physical, and moral support when being pushed around in the system, somewhere to meet for recreation, and people one can rule with in the case of a sovereign alliance.

Different Alliances

Apart from the different types of alliances, in the case of war alliances there are also different patterns of alliance structure into which the society can be divided. What is absent from these different alliances being currently discussed is that there are no antagonistic powers forming a core of alliances as we saw in the cold war era between the USA and the Soviet Union. In these non-war alliances that I am referring to, it is true that many alliances may be advertising for more members in order to be stronger in the job markets, there are no struggles involved to win members that can make a particular alliance powerful to fight one another. The union alliances are often in battle with the purpose of fighting for wage increases and common pension schemes. In other words, the benefits and protection a union alliance gives to the *subjectus* entails a lot and that is why it is necessary that no *subjectus* should remain neutral from these types of alliances. A classification of alliances is between the *hierarchic alliances* as against the *egalitarian alliance*. A completely hierarchic alliance is one in which the leader of the alliance virtually tells the rest what to do (e.g., sovereign alliance); in a completely egalitarian alliance there is no discernible leader (e.g., marriage alliance). One can distinguish between *alliance structure* and *alliance pattern*. The alliance structure is defined as a breakdown of the total number of different smaller organisations that form an alliance into a specified number of members in each alliance (e.g., the sovereign alliances). Thus, one possible alliance structure of *ten* different smaller organisations will be one alliance of four members and two of three members. Alliance pattern, on the other hand, takes into account not only the structure but also the membership of each alliance.

1. Professional Alliances

If the *subjectus* has a profession of his own, it will be a good idea to be in alliance with his professional unions or groups. Primary and Middle School Teachers Unions, University Teachers Unions, Lawyers Unions, Nurses Unions, Medical Doctors Unions, and all other numerous professions have a union, which as a *subjectus* one can become a member of them. In developed countries, these are well organised and have different sources of information that stipulate how much one has to pay monthly or quarterly (i.e., every four months), or even annually, in order to become a member. Where there has not been enough information provided on these professional unions, there must be some form of contact where they can be reached or contacted. The latter situation is usually the case with Third World countries, where many organisations do not have proper written documents and rules guiding them.
        Some alliances offer training and education for its members in order to help them go to the battle fairly on the job market. They conduct extensive researches to find out appropriate ways of how to deal with the problem of discrimination. Recently, the Swedish University Teachers invested a huge sum of money for the purpose of getting its members to get back to the job market. They tested hypotheses to see what are the best alternatives they can provide for its members who are beset with unemployment?

2. Marriage Alliances

The marriage alliance has been discussed elsewhere in this book in detail, so we shall not say much here. It is the most popular form of alliance that is also powerful in any society one travels to. While making the decision to be part of this alliance, the *subjectus* should consider the background of his spouse, his profession, and whether the spouse is working or not. These background variables are indicators that show that the man or women the *subjectus* wants to attach himself/herself to has contacts and could be able to offer help when needed. The marriage alliance can cross boundaries, and that is why many people make it their utmost aims to secure spouses in the early stages of their stay. Some people have become successful in becoming Princes and Princesses through the marriage alliance. The marriage alliance is *egalitarian alliance* where there is no discernible leader.

3. Underground Alliances (Underground groups)

Though this is not the alliance that I shall encourage the *subjectus* to join, it has been able to help a number of people that have a strong attachment with them. In some societies, this sort of alliance has strong contact with the sovereign of the society and so can get on well in the society, more so than many of the alliances that have been described above. They hold certain key positions in the country and when they set out to harass the *subjectuses* it can be very difficult for the latter to survive in the society. This is one of the most popular organisations in the Western world and since they work underground people do not usually talk openly about them and their activities for fear of reprisals.

4. Sovereign Alliances

The easiest way to become part of the alliance of the sovereign is to join a political party in the country of domicile. There are no short cuts to these alliances, but marrying into a rich family that has influence in the country can quicken one's access to some top positions in the country. Or, to become a celebrity overnight can allow someone to join a sovereign alliance. One can receive enormous help for being a member of the sovereign alliance, though in some countries it does not entail any help. Examples of sovereign alliances apart from political parties membership are University Professors Unions, Vice Chancellor of the Universities, Company Directors, Bank Directors, etc. The sovereign alliance can be classified as completely *hierarchic* in that the leaders at the top virtually inform the rest of the minor leaderships what to do, and this makes it sometimes very uncomfortable to deal with them. The alliance is a bit loose when it concerns a native of the country, but for the *subjectus*, evidence suggest that it does not matter whether your interests are being given attention to or not.

5. Immigratory Alliances

These immigrant group alliances are basically formed to give protective support to many *subjectuses* that live in a particular region. It enables the latter to have strong contact with their roots. These can be considered as friendly alliance, especially for the *subjectus* that joins an alliance that consists of members from his native country. Evidence suggest that many individuals feel more at home to be part of these alliances than any of the rest, especially where the *subjectus* has no professional background to enable him/her become a member of a professional alliance. There have been occasions where the *subjectuses* have been well defended due to the help one had received from these friendship and interest alliances. There are many activities that certain alliances organise for its members to make them feel enjoyable to belong to such an alliance. Formal and informal agreements usually favour the *subjectuses*, and the employment of native languages at both local and official meeting means that no member is left out in any discussion or official business. Examples are the Black Moslem Group, Armenian Group, Jewish Immigrant Group, Black African Group, etc.

6. *Voluntarius* Alliances

These are alliances that one can make with the voluntary organisations. They are very popular organisations that include attachment to religious organisations, congregations, and cult groups that are found in many urban areas of the country of domicile. Though many individuals that become partners in these alliances have a primary aim of obtaining spiritual support and fellowship, evidence suggests that these alliances are more powerful in terms of the manner they defend their members in case they fall into trouble. In many cases, unless it is something that goes against the principles that are laid down by the governments as crime and the organisation as sin, members receive enormous physical and spiritual support from its alliances. Defence can be that they themselves appoint a lawyer for one or a member who is a lawyer can volunteer to defend one. These alliances have not only helped individual members in need, they usually contribute to help many political alliances; that is, if a member wants to run for political position they do their utmost best to help win the election. Examples are the Catholic Church in USA, Black Baptist Church in the Southern USA, the Red Cross, the Salvation Army, Scientology, the Mormons, etc.

7. Multiple Alliances

There are no hard and fast rules regarding the joining of several alliances, as this can increase the *subjectus's* chances of obtaining what he is after quickly. The only problems are seen to be the membership fees and time at one's disposal. Since each alliance demands something from its member; one should contribute help in the form of money, as well as have free time to attend the different meetings organised by the alliance organisations. Multiple alliances are one way of becoming popular, and it is the best approach to becoming a well known politician. Examples are to be a member of the Professors Union and a Politician at the same time, or a Secret Society member and a Priest of a strong church or denomination at the same time.

The Problem of Neutrality

The *subjectus* who wants to function normally in such societies that are so complex as many Western societies have, to remain neutral is not the best idea at all. Neutrality from alliance means one is a non-functional being unless one wants to live as a sicklier or idler. Even to open up a company which makes it possible for the *subjectus* to become independent and does not need to depend on someone else or the government for income entails some alliance, otherwise one may find it hard in getting the products that one wants. As an individual, to remain neutrality in the world of alliances means that the individual is on a separate Island that is in the middle of an ocean.

The Process of Defence, Getting Help or Bargaining for an Alliance Partner

The normal way of getting useful help from one's alliance partners in case of a legal case is to contact them through telephone, the Internet, or through a letter. If the case is such that the member has to contact them personally at their central office, the former will be directed to do so. But prior to this the member has already discussed it on the telephone, and they have agreed that his presence is needed. As this is not war whereby arms and big preparations have to be made, the member would have to wait upon the direction of his alliance partners and their lawyers. Apart from the cases when these lawyers have to defend a member, there are other cases such as increase in wages and promotions where the member in question can be represented by its alliance partners. But it can be said that apart from these numerous aids the member receives from his partners, they take care of disputes involving discrimination seriously. An attempt to find some mutually acceptable compromise will be made, and a number of different possibilities are readily suggested to both parties. In case of any major conflicts, the alliance partners are the best organs to combat for a member and also seek for an appropriate settlement that usually suits both parties.

The usual process in case of a conflict or any major crises commences with an *evaluation* of the case or the complaint that has been brought before the alliance group. Evaluations concern whether the case is something that is in their jurisdiction and appropriate to pursue. Having evaluated the case, a *decision process* is commenced and reached that will determine what is to be done next. This will simply imply whether the case should be pursued or be given up. In the latter situation this can suggest that the alliance partners have not found sufficient reason to go ahead with the case. In the former state, *an action* will then be taken which is a clear indication that the alliance partners have found sufficient reason to take up the case and defend its member or bargain in case of discrimination, a transaction, or a promotion in work place. A *favourable* or *unfavourable* response is later communicated to the member that even in the case of the latter ensuing, compensation could be paid to the alliance member.

Empirical Evidence: The Growth of Alliances

Alliance as a General Form of Protection

The growth of alliances in the Western world has been studied for many years, and empirically alliances have been found to yield positive results and benefits to all the members that join them. They provide an appropriate protection for its members and also battle for them in case of any *legal conflicts*. Economically it is safe to join alliances of one's profession but research has indicated that all the alliances mentioned above seem to offer something beneficial to its members. Even the marriage alliance, which entails just one single individual partner and his/her family members, that is the extended one, namely brothers, sisters, cousins, grandmothers and fathers, seem to provide a much better protection to the *subjectus*. There is a superficial presumption that an alliance, particularly for warlike purposes, will grow as large as possible because the larger an alliance is, the more powerful it will seem. But in the case of multiple alliances for the foreign organism there is no guarantee that belonging to many alliances will make one powerful if the individual does not make an attempt himself to do something about his situation. Since it is the needs of the *subjectus* that compels him/her to join some of these alliances, the individual must assess his needs prior to becoming a member. Multiple alliances provide the contacts that help the *subjectus* that is dynamic or charismatic to work his way through to gain what he is after. The obvious implication of this is that an alliance will not always strive for the *subjectus*. Multiple alliances are not necessarily the key to the solution of one's problems. For convenience sake, it will be asserted that the order of joining an alliance is a good thing. There are enormous benefits that will accrue from the joining of an alliance, but also there are costs involved. But certainly the benefits outweigh the costs. Judging from past situations of many *subjectuses* in the Western world, it will be unwise to remain neutral from joining alliances.

      In some alliances the bond between the *subjectus* and its partners are measured not to be strong, though the latter may have entered into partnership properly and have signed all the agreements binding them. Belonging to a particular race can affect the recognition of the partnership. This makes an individual susceptible to all sorts of problems, because it can affect the defence one receives in case of discrimination, or any other case involving the right to be defended and attended to. This means that an individual becomes disillusioned and commences having bad thoughts about the alliance and their pledge to defend when one is in trouble or having predicaments. This can easily occur because an unscrupulous person who has power wants to use the strong contacts to make things very difficult for the alliance to defend a member. There have been numerous occasions where some *subjectuses* felt their alliance partners had betrayed them. Some alliances have many other functions to play, so the fact that one has experienced dissatisfaction with his partners should not give anyone the course to withdraw his membership from the alliance.

Scholium

To become a member of an alliance is a good thing that needs to be encouraged. It opens doors for contacts that may be very good use for any person serious about getting established in a society. It offers an individual somewhere to spend time discussing, not about conflicts, but a place with people who share common interests with you and cherish your presence.

      I have provided the difference types of alliances that are open to individuals who are serious and want to become members. Those who are genuinely interested in them should make enquiries in the tourist office in the cities or book an appointment and meet with the Labour Office in your city of residence. Allegiance to these member organisations will offer greater contributions to you if you decide to become a member. Though there are some responsibilities that need to be fulfilled, the individual will soon discover that as a new member of an alliance, the benefits outweigh the time and money that will be invested in them. Let me conclude by offering some axioms or propositions that these alliance assumptions offer us. Though I move on later to present some theorems that are deduced from these axioms, the latter could as well be regarded as theorems if I were to argue them to be so.

Definitions

Definition 1

By power I understand the ability to influence or persuade someone, authority, or an organisation to do something in one's favour in order to accomplish a task. By this same reasoning, we can think of an influential group, a person, or an organisation. It involves the capacity of exerting force or doing a particular job that portrays that one can be reckoned with.

Definition 2

Protection is the act or an instance of shielding someone/ a person from difficulty or danger. It denotes here the act of keeping safety, defending, and guarding this individual.

Definition 3

By solidarity I comprehend the unity, agreement of feeling, or action, especially among individuals with common interest. Solidarity ensures mutual dependence on one another.

Axiom or Propositions of Alliances

Problem. *It is required to investigate shortly to what extent alliance formation functions as deterrence from discrimination in a society.*

Axiom 1

Whenever individuals by virtue of interest and wants, gather themselves to form social cohesion in the form of alliances, they are capable of possessing power that makes their presence felt; and by doing so, serve the interest of its strongly knit social bonds between members as well as provide them protection, which augment a hindrance for some of them from being maltreated.

Proposition 1
Theorem 1

The manners in which people are made to form alliances with some organisations and among themselves portray that there is some strength in this cohesion that is equally related to the formation of group solidarity.

It is this strength in solidarity that compels people to join the alliance partnership. Therefore, these people that join the alliance are conscious and they stick together as one force that can be reckoned with.

Proposition 2
Theorem 2

Everybody that forms alliances with some organisations possesses some kind of protection that is given to him by the group solidarity.

There is expectation from these members that these organisations that they are in bond with shall offer them some kind of protection. Without this expectation, it is possible that some people would not be eager to become members in the first place.

Proposition 3
Theorem 3

The alliance formation per say may not necessarily aid every individual in acquiring his numerous expectant wants, but it may be a bridge to creating essential contacts in the world of influences.

Proposition 4
Theorem 4

If several people are in multiple alliances at a particular point in time, I say that the one in marriage alliance who has contacts will be readily accepted and acquire a position in the society depending on the status of his spouse's relatives.

Proposition 5
Theorem 5

Under the same supposition as propositions 2 and 4, I say that everybody who joins alliances expects to be protected by these organisations in question.

There is a sense in this theorem which all will agree. People that invest interest or time in something usually expect to have something back. There is certainly some expectation of fulfilment, reward, or protection that members expect to have from their alliance partners. Even as enshrined in a marriage ceremony, a spouse ought to love his/her partner; and this is the expectation that people who come to be united as partners look forward to. That is, to be loved and cherished by the other party.

Proposition 6
Theorem 6

Supposing that the size of membership is proportional to the degree of strength and the extent of protection offered to interested alliance members themselves, I say that the professional alliance partnerships are the ones that will offer maximum protection to its alliance members rather than say marriage alliance or underground alliance groups.

For if the size of membership has association with the extent of protection an individual member receives, then it can readily be established that the professional alliance will offer the maximum protection. It is clear from the principle of alliance that many of the professional alliances have many members due to the services they offer to its enlightened members. They have rules and guidelines that make their alliance very attractive. Moreover, they offer benefits that entice members who have joined their alliance for many years. It is therefore not out of place that the membership size is larger than the underground group alliance, the sovereign alliance, the migratory alliance, etc.

Major and Minor Theorems

Theorem 7

Alliance partnership formation possesses some power that enables people to depend on them for protection.

This follows at once from propositions 1 and 2. No one doubts this major function of alliance formation, although they are not explicitly stated in their brochures. For the fact that they are capable of holding tough decisions that makes authorities listen to them and grant their request, they do constitute a form of power in the society. That is one of the major reasons why people join and surrender themselves, as well as put their trust in their power of protecting them. To negotiate for high pays or wages means one has authority of some kind that makes other governing bodies listen to one's pleas.

       Corollary 1. Therefore members depend on the alliance partners because they can be trusted.

Theorem 8

The marriage alliance possesses itself some influences and power that offer protection and this is related to the status of the spouse one is in alliance with.

This follows from propositions 1, 2, 3 and 4. This is an old institution and it has functioned for people since time immemorial when they move to a new society. When an individual is in marriage alliance with a spouse who hails from a good home with much influence, and their family or she has high status in the society, they are capable of manoeuvring for this person. Many doors of opportunity are open to this individual, and it does not take long before he establishes well in the society.

Theorem 9

Given what has been asserted and proved, interest in the alliance partners themselves is one major reason why individuals affiliate themselves with alliance partners that makes protection to be secondary.

This follows propositions 5 and 6. There is strong evidence regarding the fact that people join these alliances for the interest they have in these organisations. The interest may concern the social role played by some of these organisations in the society. People that want to get close to their own people or friends from the same country of origin, for instance, join the immigratory alliances. The underground groups have special reasons, which one of them may be due to the long interest they have developed for one another. Though they do not rule out the idea of protection.

       Corollary 1. Interest in alliance partners may be one of the reasons why some people join the alliances in the first place.

Theorem 10

All alliance partners have deep interest in its members, and they do offer some protection to the latter that reduces discrimination or hostilities in work places.

This theorem follows all the propositions or axioms of the alliance assumptions. There is no question about the fact that alliance partners have deep interest in their members. In the first place they do obtain membership fees that help run the activities of these organisations, and these are contributed from alliance members. The major protection the alliance partners offer to its members that are in close contact are in the form of protection from maltreatment in connection with work places and in the society in general. Some of the partners of the professional alliances not only represent their members in pay negotiations, but also concerning bad treatment in connection with sacking from one's job. Their presence alone serves as a prevention technique against that of any maltreatment to its members.

      Corollary 1. If alliance partners are interested in its members, then alliance is one way of obtaining security.

      Corollary 2. Hence discrimination and other serious maltreatment are not habitually reported among those that are in alliance with strong alliance partners.

      Corollary 3. Therefore alliance provides belonging.

      Corollary 4. If alliance offers a better way of controlling one's life, then alliance is progress.

Theorem 11

If people are in alliance, they spend some time together and discuss matters regarding their common interest; this make them become a force to be reckoned with as they are no longer single individuals but a strong group with influence.

Theorem 12

Let it be granted that MA is an individual who has no form of alliance, but SD has alliance partner with a known group in high places. If MA and SD were to encounter problems which bring them face to face with the Law Enforcement, SD will be in a much better position to defend himself due to his contact with this forceful alliance which may influence the outcome of the case.

Theorem 13

When individuals surrender themselves to a group that form alliances with them, they expect these group to protect them, and the alliance count their co-operation as important.

Theorem 14

The world is filled with several alliances and closely knit groups whose power enable them to have control and steer the affairs of the world.

Theorem 15

If there be two similar alliance groups AFC and afc, and AFC happens to have more funds at its disposal than afc, the one with more money/funds, that is, AFC will have much influence and would be in a better position to protect its members as well.

Theorem 16

Given the sizes of two alliance partners, it is possible to predict that the proportion of its members will be related to its access to bigger funds, which will augment its influence and power in any given society.

Theorem 17

If P, P' be two independent alliances and D, E their leadership positions respectively, the power and influence of the leadership D, E will be determined by the sizes of P, P' memberships of these alliances and their access to funds.

Theorem 18

If ABC is an alliance partner, and abc another alliance of similar kind but this alliance remains in the underground sphere, then abc has a greater probability/tendency to engage in criminal activities than ABC which is open and transparent.

Theorem 19

If AB, CA be a sovereign alliance and a voluntaris alliance respectively that also function in the public sphere, and AD, CD a subset of the sovereign alliance and a voluntaris alliance respectively that works in the underground, then the probability that AD will engage in complex transmission is greater than CD in any given society.

Theorem 20

Two alliances FA, FC whether of sovereign alliances or voluntaris alliances respectively require additional funding which will be proportional to the size of its members.

Theorem 21

If two alliances, BA and CF be a sovereign alliance and a professional alliances respectively, and A and F be their leaderships positions respectively, then A will be highly regarded with more honour and respect than F.

Theorem 22

If two big alliances ZH and PO be the underground alliance and a voluntaris alliance respectively, and H and O be their leadership positions respectively, then leader H has a higher probability to have connection with a sovereign power or ruler than leader O.

Scholium

Since the alliance partners know the importance its members are to them, they do their utmost best to protect them as well as offer them their numerous needs. However, there are some alliances that naturally care for its members not because they ought to do that but because they have a special bond between them that concerns love. Since time immemorial the marriage alliance has provided care, love, and extreme support for its spouses who they come into alliance with. It is one of the most effective manners in getting established in the new system that an individual comes into contact with. I submit these theorems because they educate us on the significant functions of the alliance system, which consist of *power, influence, care,* and *protection*. These rather serve the interest of the member as strategy for prevention, as well as help in conflict management much more than as forces that fight for conflict settlements.

Chapter 8: Of Climatic Law and Human Behaviour: Nature and Its Peculiarities

The Theory of Sub-races: Fables

"Art goes yet further, imitating that Rational and most excellent work of Nature, Man."

Thomas Hobbes

A host of reasons accounts for why people encounter discrimination in many societies, but in the case of discrimination among many minorities in certain European countries, it is the colour of the skin. That the colour of the skin makes (contributes to) the human organism face discrimination is not a new story. The situation became worsened when the black people of Africa fell into the hands of slave traders and suffered several generations of slavery with hard labour in the plantations of North and South America. Since the abolishing of slavery, the discrimination of people due to colour has not been intensified, though it still permeates all sectors of the modern society. The colonial governments did some things that also intensified these negative attitudes towards black people; since they allowed their researchers who were mostly anthropologists to come up with several myths which, due to the customs of Africans they were capable of supporting, hate and assigned other races as inferior.

In this modern society some learned and intelligent scholars who have left us so many years ago remind us of the myths and fables they left behind about human beings that make some people use them as weapons for discrimination. Some of these well-respected men were biologists, anthropologists, botanists, and etc., and had left their legacy behind, which some unlettered individuals and still others mentally disturbed employ to support what I consider as "delinquency" behaviour regarding hate against other races. They were men of high academic standing that today receive honour and veneration from its enthusiastic followers. Their findings continue to baffle us, and one wonders whether all the volumes they have written are true or authentic? I do not dispute with them, but could it be asserted that probably there is more to it than what we visualise in their texts today? Could we ourselves discover more facts that could provide us a better understanding or a different picture than what we have already been provided?

The biblical mythology of Noah and his three sons, Shem, Ham and Japheth (Gen. 6: 1 ff.) has been used to support the canon of discrimination for centuries. This story was even used to champion the slavery that took place in the Americas. In the story, the world was filled with sin and excessive violence, and so God commanded Noah, the righteous man among the people of his day to make a ship that would save them. In the end Noah and his three sons and their families were saved while the mythological flood engulfed the poor old fellows in the whole world. On one occasion Noah, the farmer, became intoxicated and lied down naked while his two sons, Shem and Japheth, had used cloaks to cover their father;

Ham, who originally saw his father's nakedness, had disregarded it and made no attempt to protect his father. While the story goes on, Ham and his son Canaan were said to have been cursed by Noah, and this curse, according to many adherents, has something to do with the curse of people with black colour who live mostly in the tropics. The sufferings of these people in the world, some say, was due to the curse when Ham betrayed his father's nakedness, and so their descendants have been the poor souls that are suffering in the world today. Can this simple biblical passage be employed to issue a doctrine that supports discrimination and racism in the present world? What truth can we learn from this utterance of the irresponsible, intoxicated Noah after his hangover? Is the text authentic that could be used to support this ideology of a curse?

First, there are several points that can be made in connection with this myth. That is, while the incident of drinking wine was probable because a farmer who had worked very hard in the farm may be in need of some drink to help him relax from his tiredness, the saying of the curse may be the brothers' own invention. But supposing the curse was indeed something that came from Noah, might we infer that the curse was divine because Noah uttered it? As far as the story goes, God was not the one who uttered the curse and, therefore, one cannot ascribe the authorship of the curse to God. This is where we have to distinguish between "Thus said the Lord" and "And Noah said." There is a big difference in the these utterances, the former that is divine ensures some seriousness and the latter an ordinary human being who had just woken up from intoxication, and in his miserable hangover began to issue a curse on his dear innocent son. So as far as I am concerned, I see no reason why this utterance of a father who had just woken from his drunkenness should be applied to these modern day human descendants.

Secondly, the story says that it was Ham that discovered his father sleeping in his tent and made no attempt to protect his father from nakedness. Therefore, it was Ham that did the deed. But the fact that during the utterance of the curse, the emphasis shifted from Ham to Canaan, his son, indicates lack of authenticity of the saying. We know that Moses was the author of the Book of Genesis, and by the time he was writing the book, the Canaanites had been conquered and their sexual liberties had offended both Moses and the Israelites (cf. Gen. 19:5; Lev. 18: 3; 6-30).

Thirdly, the mention of Canaan being a slave was understandable, as the Israelites had just been liberated from four hundred years of slavery and as a result they considered any other people that might be living around as slaves. The land of Canaanites was to be inherited by the Israelites, and as reasonable human being and a statesman, to justify the cause of God and their own *selfish* cause, they must create a perfect reason why the Canaanites had to forfeit their own God-given Land. If they had been cursed to become slaves then their licentious behaviour and their destruction from the face of the earth or their being forced to become slaves was justified.

Fourthly, supposing Noah indeed said the curse to his son Ham, because the former was tired and had just woken up from his drunkenness with a hangover in full swing, there is no justification on his side to mean that because he was a holy man his utterance would be effective. In other words, the curse only works so far as Ham and his children lived. And if it was the Canaanites he was referring to in this biblical passage, these people were utterly destroyed or killed by the Israelites during the invasion of their land (cf. Joshua and Judges).

The connection of this passage to signify that people of colour and for that matter some human descendants who are suffering today in this modern world have come about because of this curse is wholly unsound. These are just *fables*, which any logically thinking human being, whether yellow or red or green, can never base his doctrine or teaching upon. Many of these stories that have been documented by the so-called prophets of God have been engineered by themselves. My criticism is not about whether God spoke to them or not, but rather because of the manner some of these happenings were presented, it is very difficult to detect which message one should assign "And God spoke." Since many of these prophets were human beings just as we are it is possible that some of the happenings (those that cannot be whole-heartedly ascribed to God) which were assigned to God had come from their own imaginations. For I find not any place that proved that God spoke to Noah concerning this curse, as he was drunk and could not have even known what he was saying. The logical relation between this saying and the empirical fact that we know is that drunkards curse frequently (e.g., tomorrow a car will knock you down, you pig!) and if you as an individual are not afraid of the curse of a drunkard you meet in the city, how much more should you be afraid of Noah's?

If one follows the story, one finds that the sons of Ham were Cush, Mizraim, Put, and Canaan. The other three descendants whose names were not mentioned by the passage all had been descendants distributed around the same area. Some of these people Moses himself had lived among when he was young. Egypt (direct descendants of Ham) built a very strong civilisation in the Nile Valley, which even today scientists find it unbelievable how they built their famous pyramids and preserved their dead kings in their tombs. The fact that the names of the other three sons were not mentioned indicates that the text was not authentic and could be a later insertion or an invention by Moses himself when he discovered how immoral the Canaanites were.

But if the principle of a superior race the theory of a curse is based on is not true, then why have some people developed more than others have, especially during the last hundred to sixty years ago? In connection with this, a ten-year-old could even empirically rightly see that differences exist between people that live in the colder climates and those in warm climates. And furthermore, the closer you get to the equator from the poles the greater the differences appear with regard to the manner people behave and also solve their pressing problems. Could this be interpreted as due to chance, or those environmental circumstances have caused these differences in people's environment and their style of living?

The Theory of Adaptability

The adaptability theory I intend to propose states that the concept Homo Sapiens refers not only to a special group of descendants of the human race, but the new breed of all descendants of other races scattered all over the earth. The premise of the adaptability theory rejects the current notion of the concept of Homo Sapiens, because it states that the latter was coined to elevate some special attributes of a particular race. In the eyes of the proponent it is a false myth which the so-called human race uses to portray their special attributes, as against the origin of all other races they look down upon and call inferior. He therefore, like Hannes Alfvén, will caution mankind, "To try to let knowledge substitute ignorance in increasingly larger regions of space and time is science."[29] What is therefore the adaptability theory?

Adaptability theory contends that the mind develops in proportion[30] to the difficulties or challenges human beings face at a particular point in time on this planet, that is, the physical world. (This I shall call the *primary law of intelligence*, to be formally known as Ayim-Aboagye's laws of intelligence with the formula *St + PθInt* . Where *PθInt* represents

---

[29] Alfvén, Hannes (1978)
[30] In Maths, proportion means equality of relationship between two sets of numbers; statement that two ratios are equal (e.g., 4 is to 8 as 6 is to 12) and 10/16 are proportional.

primary intelligence; *st* represents challenging environment.[31] The formula for the second Axiom, that is, where intelligence displaced is written like thus: ***Dp = - (PθInt + SθInt)***, where ***dp*** represents displacement, ***PθInt*** is primary intelligence, and ***SθInt*** is secondary intelligence.[32] This formula shows that at the less demanding environment, the individual has no access to both the primary intelligence and the secondary intelligence. This is to distinguish it from the ***secondary law of intelligence***, which I have termed the provision of knowledge in the rudiments of science). Thus the greater the magnitude of problems a particular group encounters in their vicinity, the higher the development of the brain accommodates to meet and solve their predicaments. *Accommodation* means an adjustment or adaptation to suit a special or different purpose. In other words, automatic adjustment as it occurs in the focus of the eye by flattening or thickening of the lens. Thus, on the contrary, the less challenges the environment presents to the organism, the poorer the development of the brain to adapt to this environment. With this goes the rationalisation that all men are endowed with some kind of inner excellence, which in Latin is called *excellentia* or *excellere* (as *ex*, *celsus* meaning 'lofty'), but the environment causes the differences we visualise in the manner people think and solve their problems. These concern a whole range of problems with regard to the development of complex forms of technology to tackle specific problems of the environment to the simple idea of having a shelter over one's head. *That a child subject who perchance is born and introduce into the cold climate of the Tundra Zone may be the first to organise well his life and begin utilising his brain to tackle his emerging problems accessible to his awareness, in fact concerning all sorts of problems inherent in the hard climatic zone. The same species of child that is delivered into the warm hospitable climate of Africa or the Middle East would not catch the **trick** that concerns organisation.*[33]

The adaptability theory does not reject Darwin's theory of evolution, but on the contrary it provides it with a new perspective. I depend mainly on Charles R. Darwin's publication of *The Origin of Species by Means of Natural Selections* in 1859 as my major data source.[34] Moreover, Georges L. L. Buffon, Jean Baptiste Pierre Antoine de Monet, Chevalier de Lamarck (1744-1829), Charles Robert Darwin (1809-1882), who was Lamarck's successor and grandfather of Charles Darwin and George Cuvier's (1769-1832) data that make me to accept the world as older than it appears in the biblical chronology. It is a useful data that as a social scientist I shall never discard. My theory is partially based on their geological materials, and especially on Sir Charles Lyell, (*Principles of Geology*, published in 1830), whose studies on geological material enabled him to formulate a theory known as ***uniformatarianism***. According to this theory, the geologic history of the earth had been slow, uniform, orderly development brought by accumulation of constant small changes. Adaptability theory preserves the unique quality of man in that it embraces the notion that states that at the preliminary stages of men they were quite near and close to animals in the manner they fought for survival over the same resources in the environment. But during a

---

[31] Alternative formula is þ= 0 x E+ where þ represents brain development; E+ represents challenging environment; 0 represents the general constant setting the rate of development. The formula for the second Axiom, that is, where intelligence displaced is written like thus: þ = -0 x E-, where E- represents the less demanding environment. *O* is called a "constant" because though the brain's physical size does not change—however its development varies from person to person and this is called "general constant".

[32] Cf the end of this chapter for a summary of these formulae.

[33] This kind of thought experiment is known in physics as a *gedanken* experiment. And for *gedanken* experiments equipment is engineered for conceptual economy.

[34] A brief history about Darwin: Darwin was educated in Edinburgh and Cambridge Universities. He was interested in natural history though he set out to study medicine and then pursue a career in the Church. He was largely self-taught in sciences and the most important part of this informal education took place between 1831 and 1836 when he sailed around the world on H. M. S. *Beagle*. The *Beagle* was dispatched by the British government on a mission of exploration and scientific discovery.

certain period of time the hordes of living organisms exposed men to competition and strife, which its positive side was the invention of new ways to adapt to the environment in order to survive. In the case of the hordes that migrated to the central part of Europe, the strivings in the harsh weather conditions meant that only those whose inner and outer dispositions aided them to produce abundant food, build good houses for protection, able to hunt animals for the skins to be used as clothes, resistant to diseases, able to later make their own original clothes, did survive. In short, the survival of the men in these harsh climatic zones did not depend so much on who was "fit "or "strong" or "the variability that produced change in the human species" as Darwin postulated, but rather those who were by nature endowed with ***craftsmanship, good wit*** to diffuse the boredom in the long winters, able to read signs of the weather by that, the first cycle of seasons were jotted down in the memory to survive till the next available warm weather and able to endure loneliness or isolation. Wit here can be considered as part of the natural wit that principally concerns (1) ***celerity of imagining*** and (2) ***steady direction to some approved end***. This can be distinguished from acquired wit, which can only be obtained through method and instruction. Those family members that migrated to parts of the well-habitable portion of the colder climate of the earth survived not because they were strong or fit alone to live continuously, but by sheer luck on their side that they were men that personally possessed extraordinary artful abilities; that is, those who nature had endowed with tremendous natural ***talents.*** This simply explains why the people in these parts of the world are seen to be hardworking men and women probably more than any other people on earth and this has come about due to the survival of only those with extraordinary talents and ***not any superior new species,*** as Darwin postulated. This is a myth, "a wonderful myth maybe, which deserves a place of honour in the columbarium which already contains the Indian myth of a cyclic Universe, the Chinese cosmic egg, the Biblical myth of creation in six days, the Ptolemaic cosmological myth, and many others."[35] In fact, the adaptability theory makes us right away rule out any superiority with special species, groups, or race as the "Homo sapiens" or the "African sapiens," "Asian sapiens," and so forth ,though it concurs with favourable variations that took place within the human species. These are inappropriate jargons and are used by "our dear brothers" to extol themselves, in which I have elsewhere designated it as ***an illusion***. Many philosophers, including Voltaire, did not regard any human being as superior in terms of the manner they are endowed with intelligence. But that special circumstances (i.e., thousands of years in isolation of the African families, American Indian families, and Amazon Indian families from the hordes) and the climatic law has made some men precedence over others. Thomas Hobbes (1588-1679) states it in his masterpiece, *Leviathan*, which was published in May 1651: "There is no other act of mans mind, that I can remember, naturally planted in him, so as to need no other thing, to the exercise of it, but to be born a man, and live with his five senses.[36] Those other faculties, ... are acquired, and increased by study and industry; and of most men learned by instruction and discipline...."[37] "For besides sense, and thoughts, and the train of thoughts, the mind of man has no other motion; though by the help of speech, and method, the same faculties may be improved to such a height as to distinguish men from all other living creatures."[38] Aristotle on his part describes this process of what I have called "discovering knowledge" in a famous passage: "Though sense-perception is innate in all... in some the sense-perception comes to persist, in others it does not." So in some people that "this persistence does not come to be, have either no knowledge at all outside the act of perceiving, or no knowledge of objects of which no impression persists...." Those "in which it does come into being have perception and can

---

[35] Alfvén, Hannes(1978) .
[36] At presents scientists have discovered that the human organism possesses more than five senses.
[37] Hobbes, Thomas (1665).
[38] *Ibid.*

continue to retain the sense-impression in the soul: when such persistence is frequently repeated a further distinction at once arises between those which out of persistence of such sense-impressions develop a power of systemitizing them and those which do not." It is from this "sense-perception comes to be what we call memory, and out of frequently repeated memories of the same thing develops experience; for a number of memories constitute a single experience." Furthermore, "from experience again—i.e., from the universal now stabilized in its entirety within the soul, the one beside the many which is a single identity within them all—originate the skill of the craftsman and the knowledge of the man...." Aristotle concludes, and states that "these states of knowledge are neither innate in a determinate form, nor developed from other higher states of knowledge, but from sense-perception. It is like a rout in battle stopped by first one man making a stand and then another, until the original formation has been restored...."[39] For as one of the greatest thinkers in history, Aristotle believes that these processes (i.e., sense-impression, sense-perception, memories, develops experience, craftsman, knowledge of man) are important stages in human acquisition of knowledge. Aristotle's theory was to be developed later by such men as Locke, Berkeley, Hartley, Condillac, and Hume. This was the line Hobbes took.[40] But for the interim it can be commented that in the case of the Africans, the recent slave trade did untold harm to them when in their pre-industrial stage of economic development years the gifted among them were transported and maltreated for more than 280 years.

Where people lived in hordes there was always fights over the scarcity of resources, and also fights for supremacy to determine the one who will become the ruler and the rest subjects. As these parts of the world were filled with all sorts of people, the wars that ensued, I do not think historians have correctly assessed them. These wars, though we read a little about them in mythologies and fables, all have not been documented, as these early years

---

[39] Cohen & Nagel (1949:274-275).
[40] Following the rejection of this Scholastic notion by Renée Descartes who later advocated the innate knowledge theory, the Age of Enlightenment saw men like Locke, Condillac, Hume and Berkeley who returned to the Scholastic idea of sensation and reflection. The two theories that dominated the Enlightenment period were the Theory of Vibration, which was advocated by David Hartley, and the Theory of Association, which was championed by Aristotle, Hobbes and Locke. David Hartley was much influenced by Isaac Newton. Familiar with *Principia* or the *Opticks*, he was to discover and establish the general laws of action, affecting the subject under consideration, from certain select, well-defined, and well-attested phenomena, and then to explain and predict the other phenomena by these laws.

    The theory of vibration advocated by Hartley was rather clear and simple. Sensations and ideas, which both "internal feelings," are presented to the mind by the "white Medullary Substance of the Brain," which, with "the spinal Marrow, and the Nerves proceeding from them," functions as the "immediate Instrument" of "Sensation and Motion". The medullary substance was conceived by Hartley as a collection of infinitesimally tiny particles that are equipped to act as transmitters of stimuli. The nerves then reply to sensations by vibrating, and yield vibrations among the medullary particles. Accordingly, "Eternal objects, being corporeal, can act upon the nerves and brain, which are also corporeal, by nothing but compressing motion on them." The vibration theory views the relation of sensation reply, that is, vibration, as constant and predictable. Moreover, as sensations alter, ideas change with them as well. However this theory had limitations.

    The theory of association, on the other hand, explained man's sense of confidence in the world. It adheres strictly to the notion that nothing is certain outside of mathematics. Accordingly to Hume, the materials of experience are sensations and no sensation can warrant the appearance of another sensation. "Yet man does see the world as an ordered pattern, and 'common subjects' of his 'thoughts and reasoning' are 'complex ideas.' Now this order and complexity, both indispensable, yet both in a sense artificial, are the products of the association of ideas. Sensations are repeated, sensations are like one another, sensations appear to have certain invariable effects, and it is from these three 'principles of union' – contiguity, resemblance, and causation—that man constructs his mental world; association alone makes possible relations, modes, complex structures—in a word, organized thinking and rational discourse." (cf. Gay, 1973:182-183). Thus in summary, the theory of vibrations explains the action of sensation in human organism, while the theory of associations, which is its twin, explains the construction of simple sensations into man's total experience. This is why the use of language that is fixed in the organism is important in the area of education. I have discussed this at length at the appendix.

were devoid of writing and only oral tradition may have enabled men to keep records of them. It was the great wars of the Ancient Babylonian Empire, Persian Empire, Grecian Empire, and the Roman Empire that we are informed properly how those wars that had gone on for so many thousands of years undocumented look like. Unlike many other parts, these isolated cases where the African families, Amazonian families, and the American Indian families lived, the wars that went on in these other parts of the Great world where the majority of people lived spread over a wide area. (By the way the different many languages of these families that lived in isolation in the different forests depict that members were not closer to each other from the beginning, which resulted in the emergence of different languages that became rampant). These meant that wars for the hordes in the well-habitable portion of the colder climate of the earth had the purpose to rule a large area and occupy while using organisational forces to control a large area conquered. The uniform culture, which we see in the well-habitable portion of the colder climate of the earth, did not only originate from the Roman Empire as we usually are inclined to believe, but that this was the characteristic with many distant wars (that were fought many thousand of years ago) that were waged, which even the biblical mythology of Babel recounted its version. According to this myth, long before the spread of mankind to other distant parts of the world (if it is only true) men were living in hordes, and they attempted to build up major projects which would probably aid them to have contact with their illusory gods that they believed created the world and them. These unsuccessful attempts led to confusion and the beginning of different languages that finally culminated in the long journeys of many families to distant parts of the corners of the earth. Note that if we were to accept this myth, then we would even save ourselves the trouble from the coining of any theory to elevate some people as Homo Sapiens and others as Asian Sapiens or African Sapiens. It is high time we caught the lies of our dear old brothers who perchance through luck and their own hardworking and good behavioural characteristics **acquired** through contact with severe hard conditions in the environment to develop and as a result lead ahead in progress or advancement. Soon the Homo Sapiens in Asia will prove to the world that no human being possesses innate superiority but that those acquired characteristics through pressure from the environment and hard industry makes a man reach higher places. If one perchance is able to win huge sums of money today, one can create a myth around one's riches such that for the rest of the next generations to come people will believe one is super in a special manner of acquiring money from winning the lottery. In other words, luck and fame can make a group of people or a nation to succeed in creating a myth about their own illusory superior qualities that they have succeeded in obtaining through hard work, cheating, or through other circumstances (e.g., selling weapons to a great nation at war in order to earn money for development). This is possible if we also imagine that the acquisition of Moses' Ten Commandments, King Hammurabi's Law Code, the fabrication of Laws by many other religious leaders of Ancient times as well as today; all these laws were formulated by men with philosophic minds, but what do we see and hear of today? The attribution of these holy laws to a supernatural God or god that they believe authored them by themselves caused fear, and people have accepted them as authentic, divine, inspirational, or special. By creating the Homo Sapiens myths, a particular race has succeeded in ruling the whole world by their cunning ways on the majority of Homo Sapiens in other parts of the world. *When one day mankind, by searching their own hearts and come to know the "truth" about the true Homo Sapiens, that is when one could say to his equal neighbour Homo Sapiens, "I know what you so and so is telling me about this product is correct, but I don't want it," or "I know your gifts or presents you are giving to me are worth millions, but as a matter fact, I don't want them. What do you say?"* This will be the GENERATION of the end of discrimination in which the MAJORITY OF ALL HOMO SAPIENS WILL RULE. With the banishment of discrimination, the poor nation or group could say, "I am poor but I say I don't need your gifts or enticements, I have brains, and I shall depend on my own wisdom and wit to provide for myself, my family and my nation."

Not that there was no evolution of species, but that the evolution did not cause or result in widespread differences in intelligence of the different races. Rather than differences in appearances of their cultures, which were based on environmental variables such as religious beliefs and its concomitant superstitions, manners of eating, and behavioural characteristics. Religion, for instance, was the monster that impeded the progress of many families that lived in the warmer areas as compared with those that inhabited the extreme harsh environment. This, as I shall term it as ***displacement***, is the tendency of organism, subjected to conflict or frustration to engage in some activity that is part of its repertoire, but apparently irrelevant to the needs of the moment. Having not many problems to challenge their minds and bodies concerning their survival, the anxiety-laden races thought and spent their enormous time on their religious beliefs and the life to come. Their preoccupation of their minds on these things made them engaged in erecting enormous temples to house their gods, their contemplating on imminent death of their mortal bodies, performance of the diverse religious rituals. These blinded their eyes, and the result was their inability to exercise innate capacity to develop intelligence to tackle mans numerous planetary problems. Natural seeds of religion such as opinion of ghosts, ignorance of second causes, devotion towards what men fear, and taking of things casual for prognostics, occupied the minds of some of these races. While they occupied their time and minds on these religious ritual, people that the severe environment challenged them already began to use their minds as their *saviour* in dealing with their own immediate pressing problems. This unconsciously enabled them to depend on their own wisdom in the early stages of man's history to tackle their domestic and immediate problems, which became a source of inspiration for them, their offspring, and great descendants. Through this they acquired special characteristics of adapting to their unfriendly environment. Otherwise, men by nature have been equal and are still equal, and have had the right to everything and many resources. The differences between men then had been strength, form (physical appearances), arts, eloquence, liberality, specific class differences to each specific culture.[41] Thus Helvétius, the extreme environmentalists, thought that education could make man into almost anything, for he observed that all men are endowed with the same bundle of potentialities. And more so David Hume in his excellent treatise revealed that "It is universally acknowledged that there is a great uniformity among the actions of men, in all nations and ages, and that human nature remains still the same, in its principles and operations. The same motives always produce the same actions." Furthermore, "ambition, avarice, self-love, vanity, friendship, generosity, public spirit," commingled "in various degrees, and distributed through society, have been, from the beginning of the world, and still are, the source of all the actions and enterprizes, which have ever been observed among mankind." Still, "Mankind are so much the same, in all times and places, that history informs us of nothing new or strange in this particular."[42]

For the Homo Sapiens of Africa, some scholars have assigned the "African Sapiens" to distinguish these people from themselves because the Africans do not think or act like they do or they see the Africans as having a different colour as distinguished from their own. I strongly reject that view, because not long ago some people in the colder climates of the earth were as ignorant as many other races (burdened with religion and superstition) until the enlightenment where individual scholars through illumination and awareness, ridiculed the

---

[41] In the Amazon region, Darwin himself noticed that "geographic variation might be explained by the evolutionary divergence of what had been originally a single homogenous population." (Garraty & Gray, 1972:29).
[42] An Enquiry Concerning Human Understanding, in Works, IV, p.68.

manners of the people and through that brought some changes in the society.[43] But why the Africans as well as the Indians remained ignorant for sometimes and also had their skin changed from white to black or maintained their original skin colour, if you please, it is still explainable by the adaptability theory. In the first place, if the skin colour was changed from white to black it is only to adapt to the environment where the ultra violet ray, radiation, and etc., caused extreme damage to the skin without the melanin in the colour. We know that animals adapt to the environment if they live there for so many thousands of years, and this same thing occurred with human beings, too. In fact, this could have occurred supposing their forefathers drifted away from the perfect climatic region of the Mediterranean; that is, if men first lived in that region from the inception. But if they originally possessed this black colour, as the current theory has postulated, then it is understandable why they should have the melanin to protect them from the sun's enervating heat. **Chemical reaction theory** states that whenever heat or something causes matter to burn, there is a loss of quantity of the matter. Others will increase in weight. It is possible that this form of heat that the sun[44] produces causes the volume of brain to shrink,[45] the colour of skin to change, texture of the human body to change as well (melanin replacing something else). Radiation that is absorbed by the earth's atmosphere alone includes X-rays, gamma rays, infrared and ultraviolet rays. Most natural bodies loose weight and colour when heat or fire burns them, for example, wood. The human body likewise is made up of carbon, oxygen, hydrogen, nitrogen, and phosphorous, together with mineral elements such as calcium, iron, and cobalt. These are *susceptible* to the burning heat of the sun.[46] While not disguising the effect of the sun on those in warm climates, I would like to say this helps us explain the external differences we have between men, which means it does not have to change the Homo Sapiens a particular group of people are or the Africans are. No matter the arguments that have been put forward by many scholars, it goes without saying that the hordes of life of the people of Europe and the fact that they were enabled to build better houses to withstand their severe weather conditions, make good clothes to protect them from dying/extinct from cold weather, and able to make tools to help them harvest enough to survive, as well as organise themselves to meet different cycles of the weather, make them consider themselves as Homo Sapiens. These are not convincing arguments at all, as I have shown that it was a matter of one either thinks or applies trial and error methods to deal with one's dire situations else one dies from the unfriendliness of the severe environment. Africans are Homo Sapiens as anyone else but the several thousand of years in *isolation* inside the thick forests with rugged mountains, the infestation of the tsetse fly in the region, and the expanse Sahara desert that acted as a barrier between the hordes and the African families, all these impeded the progress of the Africans.[47] Where many different

---

[43] Sometime in the eleventh century, the belief rapidly spread across Europe that kings, blessed in some mysterious sense with divine attributes, were more than ordinary physicians: their touch was reputed to cure a variety of diseases, especially scrofula, "the king's evil." The king's touch soon became, and long remained an awesome and cherished ceremony, and it survived into the age of absolutism as part of the baggage of royal prestige (Gay, 1969:27).

[44] The sun burns its *hydrogen* to *helium* and if it continues it will result in *carbon*, which will later become *oxygen*. This transmutation of hydrogen into helium produces energy for the sun and the other stars. It has long been established that extremely high-energy radiation, either photons or particles bombard the earth.

[45] The theory that explains that bigger brains produce higher intelligence has been disproved by the fact that Einstein and a few other geniuses had smaller brains as compared with many others. Recent theory explains it that the size of the brains matters little as the brain with its rubber-like texture infates depending on the amount of knowledge and experiences acquired by the organism.

[46] Ponder over Sir Isaac Newton's question: Qu. 6. "Do not black bodies conceive heat more easily from light than those of other colours do, by reason that the light falling on them is not reflected outwards, but enters the bodies, and is often reflected and refracted within them, until it be stifled and lost?"(Hutchins, 1952:516).

[47] Some authors simply explain it that "almost all the great rivers of Africa descend to the sea via rapids and waterfalls, and thus fail to provide an easy means of communication (such as all other continents possess) from

cultures meet, even simple mistakes that are made by some members of them can originate good ideas and development. This can be said to be the isolation that occurred between the people in the Americas and other parts of the world. It may well explain why the northern territories of the planet earth's development came late after the First and Second World Wars, because the environment was isolated and as hostile as one can think of and they themselves needed impetus from others before they could reach their goal. At last when the Africans were exposed to people from different continents it was not a fair trade, but "persuasion to allow me take one of you to see where I live or come from" which commenced the evil idea of slavery. I presume the idea of slavery, though the African was profoundly ignorant (as many races that live in isolation for a long time become ignorant, including people from the other parts of the earth which the philosopher David Hume in 1700 also talked about or referred to some of them) did not come to the African in the beginning the way we see and understand it today. It was due to kindness that many pioneering African families offered to allow their relatives or themselves to join the different intruders to their strange land. When they got to their destination, the story was different; either you become my servant or you would die, or I would shoot you. Today many people from the so-called civilised world use these tactics to entice innocent women and young girls from the poorly developed parts of the world to keep them in slavery to perform any function their captors at last want them to do. In short, the central elements of the theory of adaptability are stated below.

AXIOMS OF THE THEORY OF ADAPTABILITY

Axiom 1

The brain develops in proportion to the hardships/challenges (strictus) the environment presents to the organism.

Axiom 2

The brain's development can be displaced or affected by the environment that is less challenging to the organism.

Axiom 3

There can be chemical reactions to the mortal bodies that can cause interference in change in the character of development.

Axiom 4

The adaptable qualities that the organism acquires are transferred to the offspring, who successively build upon them.

Axiom 5

---

the coast to the interior. Nor can the coast itself be said to welcome the overseas mariner; much of Africa is fringed by mangrove svamp and sand bar, there are few natural harbors along its shores, and on the Atlantic side the surf is heavy." (Garraty & Gay, 1984:299).

These adaptable qualities rule out any special superior qualities inherent in the organism, but instead stresses on special talents.

Axiom 6

Talents are seen to be present in all cultures.

Axiom 7

Homo Sapiens refer to the new breeds of all races spread out in the different continents of the planet earth.

Scholium

The analysis above offers us some sounding propositions, which will be expatiated on later in our full-fledged development of the theory of adaptability. We do realise here that inborn intelligence is a characteristic of all human beings. This can be further developed by human beings depending on the conditions of the environment and the negative or positive pressures these conditions exert on the organisms. This intelligence can be displaced or affected depending on the challenges the environment offers to the organism. It also goes without saying that those chemical reactions in the form of heat or fire on the organism causes the organism to loose volume, colour, and texture of ingredients, but does not in any way make us call a piece of wood "another name." A piece of wood exposed to fire is still wood, though probably deformed in some way. *It only becomes ashes when its state of wood is extinct to become ashes, that is totally different from its first property in texture and appearance.* In a similar manner an insect becomes a butterfly, that is, has changed to entirely a new kind in texture and appearance, and also limestone becomes cement, which is a new substance in texture and appearance. The new men, according to my theory, are the descendants of all races in its present form or order.

Some Implications of the Theory of Adaptability

Darwin's theory deals with "chance but heritable variation" and the "natural selection, which moulds this process of adaptation of the species". These are the first and second cornerstones of Darwinism. **The theory of adaptability proposes the determination of species intelligence level due to specific environmental constraints or difficulty.** The implication here offers us one essential conclusion that we can make of all this data. That is, if the mind develops in dimensions to the challenges and the circumstances the environment presents to all human beings. And in a person's milieu there are no such challenges to make him strive to expand his mind as well as his outside human abilities. But it has been established that there is inner excellence in all humans, then if he has not been capable of developing these capabilities, he still possesses his inner worth and beauty which makes him a complete human being.[48] This person does not have to show the world his human abilities, there is satisfaction that this knowledge of inner worth and beauty provides him.

---

[48] In saying this I do not want to revert to the Cartesian rationalist psychology that advocated the theory of innate knowledge or intelligence. Far from it. I only want to stress the knowledge or awareness of one being capable of functioning as human being.

My criticism concerning Darwin's work is that the researcher was more concerned with the differences between the races that through his travels to the distant parts of the planet he had encountered.[49] There was also the missionaries' use of the false Biblical doctrine to support slavery among those they considered as inferior races.[50] In other words, though Darwin was a scholar yet he had been conversant with these false ideologies about superior races that were being spread around in those early centuries of expeditions, discoveries and domination. Those centuries were periods where many people, both Christians and non-Christians, were *ignorant*, even including those learned intellectuals of Darwin's day. He could, for instance, have wondered about the inhabitants he had met in some of the numerous Islands he had heard about and some of them he had visited. The fact that by European standard he saw them as not properly clothed, cultured, no form of education (by the way, some have not even heard of the invention of letters), and had even no proper houses to live in. Moreover, some of these races had no form of organisation that should be regarded as appropriate for society building (but as for this, the Romans also found the Britons and some folks in the German forest during their colonisation as barbarians, naked barbarians and uncultured, some with no clothes on and scarcely no form of education). Why did other groups in Europe dress properly, live in houses, could read and write and act in different manners that as compared to these inferior races he found a different behaviour? The idea of a superior race (species) which the Bible itself had condoned those days meant that he should find a better way to defend this in another style of approach as a scientist than the usual approach which these missionaries and others were propagating. By doing this, he succeeded in the engineering of the original idea of "superior species," which even the Greeks (the originators of modern science) and the Romans (the colonial masters of many Modern European countries) did not refer to themselves. Later, Darwin was to help in the coining of the name Homo Sapiens, which in Latin is *Sapient,* meaning wise or wisdom. For sapience, which is from the Latin word, *Sapientia*, means science. It can be distinguished from prudence (*Prudentia*), which denotes knowledge obtained from experience. Today, just as we know that the melanin causes the black colour that is characteristic with the people in the warmer climates, amino acid also makes the lighter skin that many Westerners possess. *In fact, "so far as is known, populations of African Blacks and Caucasian whites all have the same genes, but differ in the frequency with which the alleles of the genes that determine skin pigmentation occur. To describe these differences as 'genetic' is commonplace, but 'allelic' would be more accurate."*[51] All these are for the purpose of adaptability to the environment, and the organism's welfare was usually the cause of these steps to guarantee nature's care for those crafty individuals that endured and survived in the hectic conditions of the planet earth.[52] Today if we were to maintain these false notions of others as "superior

---

[49] Darwin spent much time ashore, especially in South America and Australia where he probably met many natives such as the Amazon Indians and the Aborigines. These were not highly developed and had many of the resemblance we can call "primitive cultures". Cf. Garraty & Gay (1972: 28-29).

[50] I would like to state that Darwin's wondering about these things were that of more disbelieve in Christianity as a divine revelation than a belief in it. Observe the way he describes his doubt: "... I had gradually come, by this time, to see that the Old Testament from its manifestly false history of the world and from its attributing to God the feelings of a revengeful tyrant, was no more to be trusted than the sacred books of the Hindoos, or the beliefs of any barbarian. The question then continually rose before my mind and would not be banished, --- is it credible that if God were now to make a revelation to the Hindoos, would he permit it to be connected with the belief in Vishnu, Siva, etc., as Christianity is connected with Old Testament. This appeared to me utterly incredible." See further quotation from Garraty & Gay (1972: 957 ff).

[51] Maddox (1998). Italic is mine.

[52] The frequency of alleles to produce pigmentation is proportional to the rate at which the sunrays are emitted in the locality/region. An Arab woman will then prefer to remain indoors in the African sun/heat. This is because the alleles frequency increases the more time spent in the African sun.

races" or "distinct," and others "inferior," it is because we ourselves are ignorant about the past and the fact that time changes, and that the times that the Romans labelled many Europeans as barbarians, naked barbarians, ignorant, and uncultured, are still with us.[53]

Imitation and the Theory of Adaptability

If we will be honest with ourselves, we shall all agree on a point that all the developments that we see today have been transferred to mankind and its descendants through the standard of "imitation," which in Latin is *imitatio*. The word implies the manner of following the example of someone in order to **make a copy of** what he had done or accomplished. Absolutely, no magic had ever taken place in the evolution of human developments. Nor were those societies considered "superior species" before they could develop. As has been asserted earlier on, talents are found in all cultures, and mankind through nature had been endowed with these "special gifts" so that human beings could contribute individually to its people and culture, and also help in the organisation and building of its societies. Not all men possess these special talents, but those that do not have talents, it does not mean that they are stupid, but instead they may be good in other vital areas. For the fact that these people lacking special talents are around to offer a helping hand to "something" that is being done is good enough. Furthermore, it must be well understood that no human being is useless, even the one who may be disturbed with temporary illness or insanity is worthwhile. Those who have been conversant with this secret have for a long time considered everything that we have around us as useful or vital for certain purposes.

Now, the reason why those who were talented lived among other people was to use them for the benefit of all. Once an idea occurred to the *talented individual* (genius), he would share it to all so that the other people living nearby or the community could also make use of those ideas. These took place quite clearly among the hordes in the early years in the world or wherever they lived. The appropriate manner that those new ideas or innovations discovered by the talented could be transferred was to "*imitate*" or "*copy*" them. It was through this that many inhabitants that crowded in one specific area could progress faster as against those who lived in isolation in the far corners of the earth. So that those who were good in imitation once they discovered that a new idea had come up would quickly imitate, and sooner or later these ideas were assimilated into their own cognition and consequently into their culture. As they could migrate freely, they succeeded in getting closer to neighbours who never thought it was a crime to imitate new ideas or ways from them. Nobody used another person's language for the purpose of imitation, since these ideas were accessible they could make them available to all that were willing to imitate and implant them in their society. By using their ***own language*** to practice these kinds of imitations, they quickly could make them their own or adapt them and through successive learning made them their own.[54] Consequently, they succeeded in bettering them to a higher level. Imitation had one great

---

[53] Concerning the Britons, Gibbon provides this information: "A hope is expressed by Pomponius Mela, 1. Iii. C. 6 (he wrote under Claudius), that, by the success of the Romans, the Island and its savages inhabitants would soon be better known. It is amusing enough to peruse such passages in the midst of London." (Gibbon 1896-1902; quoted in Gay 1973: 624). " The masters of the fairest and most wealthy climates of the globe turned with contempt from gloomy hills assailed by the winter tempest, from lakes concealed in a blue mist, and from cold and lonely heaths, over which the dear of the forest were chased by a troop of naked barbarians." (*Ibid.*, p. 625). "The forests and morasses of Germany were filled with a hardy race of barbarians, who despised life when it was separated from freedom; ...." (Gibbon 1896-1902; quoted in Gay, 1973:623).

[54] Many psychologists will agree that consciousness is essentially the production of new behaviours, new interactions, and new relations. Learning to ride a bike or a car is a highly conscious action; it becomes entirely unconscious once it is automatic and unchanging.

advantage. As the organism imitated that idea or innovation, a thing that was not his own, the mind began to adapt in such a manner that a new phase of development occurred whereby the organism was able to perform or create a style that was different from what the organism through ignorance imitated from the talented. In the end, the novice organism became original in his imitation of the innovations and even went further to perform them in another peculiar manner that was totally different from what he imitated. And as for this we are indebted to the behaviour theorists, who have also talked about **acquisition of standard through modelling.** But still observe that there are substantial cultural and individual differences in what people consider as satisfactory in the obtaining of standards. While others follow stringent standards of conduct and reward themselves sparingly, others are quite self-satisfied with ordinary or even mediocre performances.

It is a fact that all other developments that took place in these early years of man's history occurred through the canon of imitation. And even today it could be argued with strength that all developments that had taken place in the world had happened through imitation, and that all the developed countries that had succeeded in development were one-time **extraordinary good imitators**. Imitation was the first stage that each society that became developed went through before becoming what they are today. Therefore, any society that intends to develop must rehearse to be good imitators, without this no nation will make any progress. Rehearsal helps the acquisition of learning and retention. Rehearsal serves as an important memory aid. Individuals who cognitively rehearse or actually perform imitation patterns of behaviour or innovation are less likely to forget them than those who neither think about them nor practice what they have seen. Many people had wondered at first when they saw "our brothers" who had succeeded in development and had all things they needed, sometimes they had attributed it to their being "superior species" over others. The point is that no man has any superior wisdom in these matters, the idea occurred to one or a group of people with talents and they were successful in "imitating" these ideas or innovations from them. There is no magic involved in acquiring these things, and so the earlier the society that is in need of commencing imitation begins to imitate, the better it would be for its people. No one should spend too much time wondering when his or her time will come, and ask not whether it is possible that this will happen. Start imitating, commence transplanting, and translate them into your own language and understanding. Make them adaptable to your way of life. If it is possible, practice it everyday and every week, and you are on the way to getting the **trick,** for there is absolutely no **magic** involved. The earlier the society in question commences imitating, the better it will be for them to get on the road to successful development. There is no shortcut or another way whereby one can develop, just imitate! Tomorrow one will become an expert, and even develop better than what one has imitated earlier. According to my own proposition, the processes in imitation can be divided into three phases: (1) **the primary frontis,** which is considered the initial phase of imitation; (2) **the secondary frontis,** where a new phase of development leads to originality of thinking and a distinct better product; and (3) **the main frontis,** which supersedes the first and second phases, gives the imitator omnipotent control over that which is being produced or developed.

The Primary *Frontis*

This is the initial phase of imitation where the organism begins to order his memory and those innovations it wants to copy. It comprises such activities as plans to secure appropriate materials, at what price, and whether the acquisition will have to involve normal agreement or observation of the products to be copied. If the plan has to encompass the stealing approach or spying, what steps are needed in order to accomplish this? Where the purchasing approach has to be followed, the organism will have to weigh whether it has to depend on other organisms or people to help get some intricate materials (either finished products or raw materials) from them. At the primary *frontis*, it is always better that the organism co-operate with other organisms who can supply basic materials to help beat down the cost of materials or save time. The products at this level are considered to be ***regular*** and produced in ***quantity***.

The Secondary *Frontis*

This is the phase where the organism has successfully imitated products or innovation. They have come to the stage where its neighbours know their products. Though they have succeeded in penetrating the market, still people are not used to their products, only those who can afford them because of its cheap prices. These cannot afford the high quality products that are expensive. Nevertheless, at this phase of secondary *frontis*, the organism has attained originality and even though people reject them due to their cheap prices in relationship to the original innovators. The significant development of this stage is that there is no difference between the imitator's products and the original innovators. There is ***originality, quality,*** and ***quantity*** in the performance dimensions. These products of the former are able to compete with the latter, and this is where conflict can ensue, especially if the selling prices of the former are less than the latter (e.g., China). The process whereby the products at the secondary *frontis* become available to consumers shall be termed *proximity*.

The Main *Frontis*

The organism has reached the omnipotent control in imitation, and this has superseded (in Latin *supersedĕre* 'be superior to') both the primary *frontis* and the secondary *frontis*. This stage has become possible through an energising principle known as the ***social referential comparisons***.[55] This, which is the appraisal of an individual or groups' own behaviour, requires comparisons among three major information sources, namely performance level, internal standards, and the performance of others. In the ***normative comparison***, standard norms based on representative groups are used to determine one's relative standing. ***Social comparison*** involves comparisons among associates, neighbouring societies, or people in other settings engaged in similar endeavours. ***Self-comparison*** is where ongoing performance is judged. It supplies the measure of adequacy. Finally, ***collective comparison*** is partly one's own behaviour taking other forms in society organised around collectivist principles. It is group performance rather than individual accomplishment that is evaluated and publicly acclaimed. But, though this phase is ideal, yet it can be the stage where the organism, because of competition with imitators at the secondary *frontis* could loose market for its perfect products. Because it has attained some measure of standard, the prices of its products have soured in relation to its ***high quality***. Since it cannot sell cheap, it may be compelled to take a soft attitude towards its competitors, otherwise it will loose market and revenue. Where it takes the unpopular steps of intimidation and threat, it may lead to disastrous conflict unless an agreement is reached where both organisms from rival countries can maintain a stable relationship in the market (e.g., Sweden, Germany, and USA).[56]

---

[55] I am indebted to Albert Bandura whose work provides these concepts.
[56] In terms of development, we can say that England was the talented that commenced the industrial revolution, and all others copied from him. The English themselves were dependent on other cultures in their breakthrough regarding the use of machines to increase productions.

Sapience as a Myth

We have established through the above analysis that the formula for progressive development in this physical world is ***imitation*** and ***adaptation*** (imitate and adapt!), which should occur in the organism's own cultural context. Now I would like us to consider the word "*sapience,*" which originally comes from the Latin word meaning "the wise" or "wisdom." It becomes apparent that the meaning of the word was probably one of the reasons that occasioned the scientific pioneers to assign it to the "new man." I have rejected it as a myth! The question that I want to pose now is, why should these pioneers designate it to this mythical new man as the wise and the rest as something else? Why should they not regard this man as simply "man" like any other race? Was it because the period they were coining this name these men found only the new man's race as the most cultured on the earth? Does the name Homo Sapiens give them a unique preference of a particular race as against other races? Why did they rank men and apportioned a particular race as the number one? Why not all races as Homo Sapiens?

       The scientific study of man and the theory of evolution have contributed a lot to mankind, especially for the fact that they opened up interest in the investigation of the human species origins.[57] Their use of fossils to establish the generation of man and its different species has been regarded as one outstanding achievement. But while it has used modern techniques to unravel the history of human species, it cannot be said for sure that their attempts have been successful. Their postulates and numerous speculations are not one-hundred percent correct. Though the techniques modern anthropologists and archaeologists are utilising are superb, there are a lot of speculations that cannot be proved scientifically, as they depend on old bones that have remained for many thousand of years. And so while they speculate so much without getting more from scientific observation, it leaves room for criticisms. It is absolutely correct that development was gradual and took many thousand of years before mankind reached this stage where we are now. The fact that modern industrial development had taken approximately 250 years or more to materialise indicates that the practices of man, which were superstitiously inclined, impeded man's progress. All the races were struggling with religion, superstition, and fear of the unknown, and their imminent mortal death. As soon as we were capable of untangling the numerous myths of religion, we made a way ahead for the development of modern civilisation, and this has consequently brought us to this technological age.

       Now, while the theorists of human evolution have done a great deal to provide us with enough information concerning man's gradual development, not everything they have come up with is scientifically correct. And as for this I shall state it categorically that though there was human evolution, the differences between races were not so much as to upgrade some as "human being" and others as "lower races."[58] Even "the absence from the fossil record of a 'missing link' between human beings and their supposed common ancestor with the Great Apes became a nineteenth-century scandal. Properly, but with unconvincing regularity, the Darwinists had no choice but to reply that the fossil record is manifestly incomplete."[59] In 1972, Gould and Niles Eldredge mounted an attack on Darwin's concept that the evolution of a species represents the accumulation of variations, just as Huxley had

---

[57] As Stephen Jay Gould has written, the development of any intelligent life is purely an accident of evolutionary theory--- as unpredictable as the development of any given species. Cf. Gould (1989).
[58] As Prigogine's work and indeed the entire history of the biosphere and the cosmos as a whole show, the development of intelligent life is but the latest phase in a long acceleration of evolution itself. Cf. Prigogine, Isabelle Stengers (1984)
[59] Maddox (1998).

earlier on criticised Darwin in that very century he published his work. These scholars marshalled evidence from the fossil record showing that it is the rule, rather than the exception that life forms remain unchanged for long periods of time, often millions of years. Gould and Eldredge termed that state of affairs *stasis*. They say there are few signs in the fossil record of the gradual changes that would betoken, on Darwin's view, the successive adaptations of species to make them ever better adjusted even to a stable environment.[60] The Biblical myths, fables, and prejudices influenced some of these scientists (i.e., Darwin and his predecessors), who investigated on the theories of evolution. This was the time when colonial and imperialist regimes were ruling the world and the case concerning slavery was at the highest scene. They were born into these **prejudicial societies** and they **grew up** and **socialised** in these societies as well. In these highly charged societies they heard of these mythological stories, and so it was not out of place when in their research they were influenced to condone with these racial prejudices that had originated from the Judaeo-Christian religion. Today many theories that are still popularised in scientific books and journals are still being criticised as not scientifically sound, because they have originated not from scientific observation but from the authority of religious myths, philosophy, or aesthetic (e.g., the Big Bang). Apart from a few scientists in the days before the enlightenment, only Descartes, Newton, Hobbes, and Bacon, whose works probably were not influenced by mythological ideas. For example, the Ptolemaic system that was based on the unquestioned acceptance of the unchanging heavens, the centrality of earth, and the necessity of perfect circular motion, was a mythical cosmology. That is all the more reason that Hannes Alfvén, Nobel Laureate 1970, believes that "To try to write a grand cosmical drama leads necessarily to myth. To try to let knowledge substitute ignorance in increasingly larger regions of space and time is science."[61] The current cosmology, he thinks, represents a return to Ptolemaic myths. "Both the Ptolemaic and Big Bang cosmology started from unquestionably correct and extremely beautiful philosophical-mathematical results. No one can study the Pythagorean science comprising the mathematical theory of music and the theory of regular polyhedra without being immensely impressed. The same holds for the theory of relativity,"[62] and the sapience myth.

     I will conclude by saying that just as Alfvén has asserted concerning many other scientific theories propounded without sufficient scientific observation, the theory of superior species or the Homo Sapience is a myth, "a wonderful myth maybe, which deserves a place of honour in the columbarium which already contains the Indian myths of a cyclic Universe, the Chinese cosmic egg, the Biblical myth of creation in six days, the Ptolemaic cosmological myth, and many others." "The difference between science and myth is the difference between critical thinking and the belief in prophets, between '*De omnibus est dubitandum*' (Everything should be questioned – Descartes) and '*Credo quia absurdum*' (I believe because it is absurd—Tertullian)."[63]

Language, Development and Continuity

---

[60] *Ibid.*
[61] Alfvén (1978).
[62] *Ibid.* Presently The Big Bang Theory is gaining a wider acceptance after two researchers at the Mount Wilson Observatory were awarded with Nobel Prize. Recent illumination from the works of Stephen Hawking, Professor at the Cambridge University is pushing this theory into the limelight.
[63] *Ibid.* According to Alfvén, "In myth, one tries to deduce how the gods must have created the world, what perfect principle must have been used."

Thus far I have been dealing with the organism's confrontation that occurred with his harsh environment, which triggered the use of the brain in a charged manner to overcome his immediate and mediate *strictus*. There can be a higher concentration of *strictus* and a lower concentration of *strictus*. **Strictus here can be defined as those critical factors that impinged on the organism's brain to compel him/her to take a crucial stance to muster all physical and mental energy to overcome his complex situation. Strictus can denote pressure; it can be likened to a cyclone that may engulf the organism if appropriate measures are not taken.** These acted as charged stimuli which in the case of the animal, it will instinctively flee from one environment to another for acute protection. But in the case of the human organism, the *strictus* naturally advises him to strike back, for he has no other choice than to deal with his frustration, fight for his survival, or succumb to death. In such occasion there is no opportunity for the brain to become **displaced, inactive, or idle, as displacement will mean death or extincion**. Besides rationalisation, there is no alternative for the organism, as he is considered trapped and has no other choice than to continue rationalising. But this early rationalisation that occurred with the organisms was devoid of proper instruction or the use of methods as these came later with education and the introduction of science. The habit of rationalising to deal with many immediate problems (better adapted housing, weather-adapted clothes, means of primitive transport, weapons for defence, cleanliness inside cold houses, activities to withstand the boredom) will later prepare the mind to adapt easily to the use of reason in the sphere of science and instruction. This was *not* the practice with those living in less challenging regions where *strictus* was not concentrated or intense (i.e., lower concentration of *strictus*). **It can therefore be hypothesised that the advantage that those in higher concentrated strictus has over against those in less concentrated strictus presently is that the former has a probable access to primary intelligence, which may or may not be the case with the latter.**

Language, as I have partially mentioned above, helped the organisms to prepare for a much better adaptation to the environment. It was not only in imitation or unguided apprenticeship that language became necessary, but also language was important for the organisms to transfer knowledge of some of the acquired characteristics; that is, those that need external communication. The use of native language gave them the necessary lead to both preserve and transfer knowledge to its offspring. Judging from the intensity of *strictus*, knowledge preservation was in a much higher standard probably almost equal to that of the use of certain basic methods and instruction. The necessity of using ones' own language is seen from the manner people who have been conquered by another nation detest the use of foreign language and regard it as a curse, for it impedes progress, original thinking, and innovation. Foreign domination not only introduces language; it uses people as slaves and labourers. This contrast could be seen that the periods of *slowest development* were in the later Bronze Age, the slave societies of 300B.C. to A. D 700, and to a lesser extent the feudal society of the Middle Ages. These societies squandered the minds of the population, reducing them to mere tools of a small ruling class. The freedom to use the organisms' own **blood** language made early development easier, as they did not have to strive to comprehend what was being transferred as knowledge through imitation.

Imitation, Talents and Adaptation

The Processes

What were the basic approaches the talented and many others that imitated utilised in acquiring their knowledge from the sources, or those that innovated or brought out some new simple techniques? How were they transferred to other people? Was there anything that portrays that there was more than extraordinary ability involved in transferring this knowledge?

The first approach was that the individual that has innovated something, be it techniques or ideas, allows others or neighbours he trust and certainly like to come and *freely (observation approach)* observe those ideas or innovations, and then copy them. This approach was very common in the early periods and it was not regarded as something that needed to be hidden. The innovator consequently became very happy as people that were allowed to come showered their blessings and admiration on him. But those that were not allowed to observe it freely, or were in a way hindered but had money, they *purchased (purchase approach)* those things themselves and then, through their common knowledge, unassembled these simple tools or machines, and through curiosity build their own tools or machines by copying them.

But where the human organism was not allowed to freely watch the workings of the tools or given the opportunity to buy those tools the next approach was to ***steal it with his senses (stealing approach)***. And in those days once a person had just walked by and seen this person using his tool or machine, the talented could steal with the eyes and keep the operation of this tool or machine in his memory. The ear was another sense organ of the organism that was employed to steal this new innovation, especially where there was no attempt to have any form of contact with the innovated thing. It should be mentioned that high concentration of the *strictus* meant that both the talented and all the people were pressured to come up with something new in order to survive. We can liken this to the time of the Grecian Empire, where people were eager/curious to hear some new stories or something new just to calm down curiosity in those days.

Though this may not have been rampant there was more probability that some organisms went as *spies (the spies approach)* in order to discover the new knowledge and then return and build their own. By this way they prevented the situation whereby they had to constantly depend on this talented individual or group, and thereby avoid becoming a servant or a slave to him or them.

There was also the *genius approach*. This person, like a dreamer, heard that someone had built a machine or tool like so and so, and in his dreams could imagine how it looked like the tool someone only described to him. He set out and built his own tool without having to have contact with the original innovator. His tool may be much better or inferior to the tool the talented had built. But that would be improved upon as time goes on.

Finally, there was the *agreement approach,* where the originator agreed to transfer this particular knowledge to other people who he wished to meet, and so he helped them and they all acquired this knowledge. The agreement approach benefited those that went and worked as apprentices. The originator was paid as part of the requirement to transfer his knowledge.

"Crises Situations" and "Aggressive Behaviour" as Necessary Familiar Notions in Adaptation

The organism's survival in the environment required the behaviour that could be well described as aggressive disposition. Here the *strictus,* as used in the manner of external pressure that was ready to engulf the organism was important. This force that was exerted from outside of the organism played an outstanding role in order for adaptation to gain a better direction. This usually took the form of **wars** or **an extreme crisis situation** that the organism was subsequently drawn into, probably not of his own accord but due to other external unexpected factors. These different kinds of tensions still appear in this modern day now and then, so for the organisms that had lived all their lives in the cold climate, this is not something new. Thus during wartime or crisis, houses would be destroyed and those valuable things the organisms had could be taken away by force or stolen. Even the farms that supplied food for the multitude of people could be destroyed. Sometimes the source of water supply could be cut off. The organism through his past knowledge had to begin from afresh and build again. This was the usual manner of *strictus,* as pressure or a force exerting on the organism was experienced or manifested, and this did not occur only once or twice, this occurred on successive periods. All these different forms of distractions had to be accommodated by the organism in the cold climate, that is, became used to how to build quickly when they were destroyed. Also how to improve on the crude weapons they had, a better manner of preserving food in times of war, and how to make better war clothes, and so on. The result of these unpredictable situations made it possible for them to deal with tensions, and also to expand their experiences, which was the direct source of *prudence*.

In those difficult times trust was not entertained for the neighbouring organisms, and if they lived as groups very distant, there was always suspicion as these enemies could make a surprise attack on them. Organisms that adapted well had to be kind to their neighbours, used the alliance principle of getting married from the other neighbouring organism's community, and being opened to themselves while maintaining secret strategies of how to defend themselves in case of war or crisis. **The Solomon code** was very commonly practised among many organisms in these warfare zones. The latter is **defined as a way whereby in times of development any groups or societies maintained some secrecy while sustaining a low profile of openness to the outside community.** It was to prevent infiltration into their private programmes that made the hordes organise themselves well to prevent any suspicion. Unity among the people was important, and a brief knowledge on what the people intended to carry out was communicated to its people. They spread that information in regard to the sacrifice they had to make if they were to survive. The sacrifice was for the future children even if they lost and could not succeed. Those foreign organisms among them should be people that are in marriage alliance; otherwise they cannot be trusted.

Usually it was in crises periods that the **talented** (geniuses) could think properly and come out with innovations and better ideas.[64] These ideas or inventions could then be worked upon during the wartime and after the war, that is, peacetime; a considerable amount of time would be spent on it to develop it better. Inventions can, therefore, be viewed as **actions** or **the products** of a crisis situation; that is, people could come out with better ideas when they are pressured during dire situations. Crisis behaviour was the one that compelled many developed nations to lead or commence major concrete developments. Therefore any land or society that intends to develop should **firstly, be very aggressive; secondly, development should be seen as the only way it can get out from modern-day slavery; thirdly, it should be seen as a challenge rather than "give us some aids," or "loans to develop," or "cure our illnesses"; and fourthly, develop a suspicious character; by this I mean to reserve trust for only few that one knows very well.** It should commence in such a way whereby people should accept that probably they should forgo all aids, humanitarian aids, loans, expect to be on their soil and say what they want to offer. The people should be conscious of infiltrators, because not all may be willing to see them get up from their dirt. Enemies prefer

---

[64] T. S. Kuhn says in his masterpiece "It is, I think, particularly in periods of acknowledged crisis that scientists have turned to philosophical analysis as a device for unlocking the riddles of their field." (Cf. Kuhn, 1970:88 ff.). We must here acknowledge the positive contributions that both scientific and political revolutions have given to inventions and developments.

that people become a slave forever so that they could point their hands at them always, "Look at those people!" There should be some *sacrifices* that should tell the people in question that "yes, we would even allow some of us to die if we can go hungry for some days before we get on our way." Any attempt to develop without becoming aggressive will not work. Politicians should spend little time in the Parliament House and instead organise those groups that can start any moment they are ready. They should start with a 10-year plan programme, and then completes the next 5 years, and they will see the difference. All the inhabitants of the land in question should be informed what the government intends to carry out and the sacrifices it wants to make, and how long the programme will take.

Box 4: Adaptable behaviour toward development

- Strict respect toward time
- Respect toward leadership and officer in charge of work
- Laws or rules must be obeyed
- Quality work instead of quantity
- Work or assignment must be *properly* done
- Transport system; Preferably Bicycles
- Group work must be taught at schools to foster team work
- High respect for yourself as an individual
- Don't beg for loans!
- High consumption of Home made products
- Nation trust yourselves and your abilities
- Be conscious about environmental pollution

Adaptability and the Theory of Science

The secondary law of intelligence is based on a single postulate:

That there is a higher degree of additional intelligence, which is the product of the utilisation of the scientific method to the scrutiny in the rudiments of the sciences.

The seventeenth-century philosopher Thomas Hobbes once wrote concerning reason and science that the light of human minds is perspicuous words, but by exact definitions first snuffed, and purged from ambiguity; the reason is the pace; increase of science, the way; and the benefit of mankind, the end.[65] I have called the provision of the mind's development and the acquisition of productive knowledge in the rudiments of science the *secondary law of intelligence*. In other words, science complemented the use of human experience, which as the source of prudence and the basic use of reason without method that was characteristic with those living in hordes regions, and it also gave them precedent over others. Symbolically we can write the formula for this secondary law as, $Steo + smt + L = S\theta Int$ where $Steo + smt$ is the study of science and the application of its methods. The L represents the study and use of the blood language.

---

[65] Hobbes (1665:36).

While we have stated the first law of the theory of adaptability as something that will happen, the second law is, after all, statistic one, stating what is likely to happen, not what ***must*** happen. Just as there is an incredibly small chance that all people that study science will become intelligent and utilise the scientific principles to solve their problems, there is even a smaller chance that all people on earth will have the opportunity to study the principles of science. The arguments concerning this second law, I believe, will not impress every scientist, since according to this theory the probability of these occurring is, in fact, so tiny that it is equivalent to impossibility. Nevertheless, we shall continue to present some arguments that indicate that the study of the principles of science have affinity with the organism adaptability to the environment, which consequently provides them certain abilities to think and possess intelligence. In the first place, science encouraged the utilisation of reason and the use of scientific methods that made the organism survive and adapt to his environment. Reason is not like sense and memory that a person is born with, nor is experience, which produces prudence, but reason is attained by method and instruction through hard industry.

The theory of science is important because it gave man the enormous power to comprehend the universe and control it. The knowledge of the natural laws put man at ease, and provided man with the impetus to further investigate the universe and make a considerable impact. The rejection of the supernatural involvement of deities in the universe made man to discover these laws that commenced the scientific investigations concerning the environment. This led to the discoveries of vast theories based on observations and empirical evidence by geologists, cosmologists, physicists, physicians, and other different branches of sciences. Theories of diseases such as widespread use of antiseptics, surgery, and the general use of vaccination were made available through science theory. Science provided the intellectual discussions of basic problems affecting the human organism; gave systematic approach to investigation of a particular problem; furnished man with critical thinking; and better organisation approach to everything in the human society. The use of *logic* helped man in a considerable way on how to reason. Logic was used to find consequences in words; adding together two names to make affirmations; to make syllogism; and many syllogisms to make a demonstration; and from the sum, or conclusion of a syllogism, subtract one proposition to find the other.

Reason has been defined as nothing but how to reckon (that is, adding and subtracting), of consequences of general names agreed upon, for the marking and signifying of man's thoughts. "The use and end of reason," according to Hobbes, "is not the finding of the sum, and truth of one, or a few consequences, remote from the first definitions, and settled signification of names; but to begin at these; and proceed from one consequence to another. For there can be no certainty of the last conclusion, without a certainty of all those affirmations and negations, on which it was grounded, and inferred."[66] Whereas sense and memory are but knowledge of fact, which is a thing past, and irrevocable; ***science is the knowledge of consequences and dependence of one fact upon another.*** By which, out of that we can presently do, we know how to do something else which we will or like another time. Because when we see how anything comes about, upon what causes, and by what manner; when the like causes come into our power, we see how to make it produce the like effects.

Before the employment of science, things were discovered through accidents and also through the trial and error methods, where an individual organism could get his problems solved. Science commenced with the asking of or questioning about everything, like "why," and always looking for a "proof" or "evidence." It was the Greeks who started with this business of looking for tangible evidence before establishing a proof of something. From Socrates, Plato, and Aristotle, these men who began to use the scientific principles were early

---

[66] *Ibid.*

thought to be crazy and accused of polluting the minds of the young men. Socrates was tried and had to die through poisoning. The new manner of perceiving reality was *not implanted in men*, it was discovered that men could question and demand a proof of everything. While we did not hear much from the Romans with regards to the methods of science except from perhaps Cicero, Carneades, Saint Aquinas, and Saint Augustine, the centuries following the Roman empire had produced important scientific discoveries by men such as Leonardo da Vinci, Kepler, Galileo, Descartes, Newton, Einstein, and Alfvén. Both the *deductive* and *inductive* approaches were employed by some of these great men to discover the workings of nature and to unveil the natural laws that govern the universe. The inventions that resulted from the industrious work of these men had consequently led to other discoveries which had led man to a stage of increase development and learning. Systematic study and the use of science methods have resulted in tremendous inventions in the seventeenth, eighteenth, and nineteenth centuries. It can be hypothesised that the brain's development of many human organisms has occurred through the use of scientific methods. Science has provided many people increased knowledge ability to reason and make good judgement. This has enabled imitation to occur without major problems in many different societies in the world.

Adaptability, Science Methods, and Self-Efficacy

Knowing what action to take in order to surmount problems in the generation of science in one's environment is vital. Scientists have asserted that this has to do with self-efficacy.[67] *Efficacy* entails a generative capability in which cognitive, social, and behavioural sub-skills must be ordered into integrated courses of action to serve incalculable purposes. Success is often attained only after generating and testing alternative forms of behaviour and strategies, which requires enormous effort and perseverance. In the ordinary world, self-doubters are quick to abort this generative process if their initial efforts prove deficient.

Scientists claim that there is a marked difference between possessing sub-skills and being able to use them well under different circumstances, as in the area of development. For this reason, different people with similar skills, or the same person on different occasions, may perform poorly, adequately, or extraordinarily. While *competent functioning* requires both skills and believes in the self-efficacy to utilise them effectively, *operative efficacy* demands continuously improvisation of multiple sub-skills to accomplish an ever-changing state of affairs. *Perceived self-efficacy* is people's judgements of the abilities to organise and execute courses of action required to attain designated types of performances. This is not concerned with the skills an individual possesses, but with judgements of what an individual can accomplish with whatever skills he has.

In the era of development, self-beliefs contribute to quality of psychosocial functioning in different manners such as *choice*, which is what courses and action to pursue; *effort expenditure and persistence*, which determines how much effort people will expend and how long they will persist in the face of obstacles or aversive experiences; *thought patterns and emotional reactions*, which is peoples' judgement of their abilities which influence their thought patterns; and finally, *humans as producers*, that is, people who see themselves as efficacious set themselves challenges that enlist their interest and involvement in activities. With regards to the last-mentioned, this makes individuals intensify their efforts when their performances fall short of their goals, make causal attributions for failures and

---

[67] Once again Albert Bandura's work provided me with this concept which has been expatiated on in his work *Social Foundations of Thought and Action. A Cognitive Theory* which was published in 1986.

support a success-orientation approach potentially threatening tasks non-anxiously, and experience little in the manner of stress reactions in demanding situations.

To become adapted in the organisms' environment meant that individuals had to develop their self-efficacy, and these developments took place in the family setting, among peer validation, schools and universities as centres of learning, transitional experiences of adolescence, through concerns of adulthood, and finally, with advancing age or maturity, with the self-efficacy of the latter, that is, the elderly centres on reappraisals and misappraisal of their capabilities. Though there is a loss in physical ability, gains in knowledge, skills, and expertise compensate for these losses.

Imitation and Scientific Contribution

The use of scientific method encourages the principle of imitation to investigate or confirm even theories already propounded by famous scientists. Presently one cannot write a scientific work without *making analysis of work* that has already been done on the topic of one's interest.[68] It is understood that knowledge progresses when they are built upon the earlier knowledge that has been accumulated in a specific area. The use of science to help bring societies ahead was practised by all those societies that are today called developed nations. Thanks to the Greeks! In some cases they assembled the talented in one particular Research Institute or College, and they tried to solve problems that could have taken years to solve were one person to attempt to decipher them. These talented men and women saw this as their duty to work together as a team in order to convey solutions to man's problems on this planet. The success of many inventions, therefore, depended on many people who together contributed to solving well-known problems that tormented the society in question.

Therefore, any society that intends to develop should first choose one Technical University, a Research Institute, Technology University, or Science College and try to bring together *only* the talented (geniuses) individuals in the society in question. This is what the Russians did; like many other countries they assembled all its scientists at one particular Institute and began their industrialisation programme. That particular university should be devoted to only these geniuses who have demonstrated that they could work in teams and be given the responsibility to work on a particular problem facing the nation in question. The processes involved in *imitation* and *adaptation* should be opened to them so that these people will learn, and try to solve the problems of the society in question. Also, the three phases of imitation: the *primary frontis* (the initial phase of imitation), the *secondary frontis* (that which leads to originality of thinking), and the *main frontis* (omnipotent control over that which is being produced). Artificial "crisis situation" could be declared or created, and all the people in the society should be alerted about the nations' intention to invent or solve a general problem that it intends to do or a programme that it wants to carry out. The nation should wear "aggressive behaviour" in its attempt to develop, as it would indicate their seriousness to carry out the programme. They should not have to wait upon promises from external financial sources, but instead should rightly go ahead with its plans and do "something"! *First of all, no nation can develop or take this initiative we are describing unless it trusts its own technicians, scholars, engineers, and architects more than any foreign professional in these areas.* Any country that trusts not its inhabitants more than any other inhabitants on earth is bound to be a failure when it considers developing. **THAT SHOULD BE THE FIRST RULE.** Never trust a foreign expert more than yours! Trust your own experts more than any

---

[68] Here again T. S. Kuhn enlightens us that "Normal research, which *is* cumulative, owes its success to the ability of scientists regularly to select problems that can be solved with conceptual and instrumental techniques close to those already in existence." (Kuhn, 1970:96).

other professional. Let foreign experts come in, but only when there is no one in your country who has no single idea about what you intend to carry out or do, before you should hire an external expert. It is better the citizen makes the mistakes and then later corrects them than just to waste money double the pay of what a novice in your society can begin and later become an expert. But though it is asserted in this way to guarantee autonomous adaptation process, this should not remove the need for external communication with experts who reside abroad. External experts' direction can be sought through the Internet and the e-mail. Imitation means that you have education concerning whatever you lay your hands on. It may not be advanced education, but still those elementary ideas will encourage these persons to learn while making their mistakes. Once the society in question has established trust for its inhabitants, be it a native citizen or the *subjectus* who has adopted citizenship, it will discourage future importation of external experts, and this will **be a clear sign that the society is on the way to real self-determination**. Any attempt to develop with the help of external experts while you have someone in the society that has knowledge in the thing being developed will not be counted as real development. It may be a counterfeit development, and the society in question will get stuck on the way to development sometime in the future. Build your self-esteems by trying to do it yourselves, that will indicate that when there are problems in connection with those developments in the future you will be able to resolve them by yourselves.

Box 5: The Talented

- **UNCONDITIONAL SURRENDER TO YOUR STATE**
- **BE AWARE AS SERVANTS TO YOUR STATE**
- **AGE OR EXPERIENCE DOES NOT MATTER**
- **STATE'S INTEREST YOUR PRIORITY**
- **TEAM WORK**
- **RESPECT & RECOGNITION OF THEIR TALENTS**

Box 6: Talents for nation building

- **SCIENTISTS**
- **ARCHITECTS**
- **ENGINEERS**
- **CARPENTERS**
- **PAINTERS**
- **SCULPTORS**
- **TECHNICIANS**
- **DOCTORS AND NURSES**

Scholium

The theory of adaptability has proposed that the human species level of intelligence could be determined by the specific constraints or difficulty the environments present. This theory has made use of Darwin's data, though as the proponent of the new theory, I do not agree with him totally on some of his conclusions regarding the derivation of superior human species. For Darwin's theory deals with "chance but heritable variation" that occurs within the species, and the "natural selection, which moulds this process of adaptation of the species." These, being the first and second cornerstones of Darwinism, are still intact but it is the origin of the superior species that I find a problem with. The adaptability theory provides us with the information that supports the fact that the differences among the species were not all that large, for apart from physical appearances, such as strength and form, men differ also in arts, eloquence, liberality, specific class differences to each specific culture. The pre-eminence of any species is based on the environmental conditions, which were termed *strictus,* and this influenced those with high concentration of *strictus* to adapt to their milieu. Those who did not catch up had their intelligence displaced or affected by their style of living. Since they had no high concentration of *strictus* to challenge their minds, they became preoccupied with other things (such as religion) that were not needed for the development of the moment.

The differences between species are marginal and should not allow us to designate some as higher than others, since the environment was the cause of these differences. The fact that there was less concentration of *strictus* in the warmer climates, and also because of chemical reaction which affected the mortal body organs, actual development of the brain, though not one-hundred percent impeded, was displaced, and so was the human progress in general. All human beings are Homo Sapiens irrespective (i.e., wise or possess wisdom) of the differences that exist today, and though others' intelligence has been displaced if it should be resuscitated could make things change and put them on the right track. Many of the developed nations did not begin very well; they were uneducated, seen as naked and barbaric, uncultured, poor, and had their fears calmed down by religion. Religion was for many centuries the ***opium*** for the human races. Yet today they have transformed themselves and their societies, and have come ahead with progress and excellent developments.

Knowledge, as we know, has been transferred to many people in the diverse societies through the canon of imitation. Through this manner of imitation and adaptation the human organism has enriched himself and his descendants, who have built successfully upon them. The success of the human races that had developed had depended on the fact that they had always and at all times educated themselves and employed their *own blood language* (the language that is fixed in the blood). The language is used from birth till death. It is by their own language they had communicated all their past history and development to its offspring. Foreign books or innovation are quickly translated into their own language and any innovation also is adapted into their own manner for consumption or use. From the first year at school, the language is the one used as a means for instruction and communication; publishing houses make sure that things needed for instruction and information are made available. In their political meetings the language used is their own language, and even in the parliament, no language is known apart from their own. Directions to different places in their cities are written in their own language, and so is information leading to what a citizen ought to know in order to enjoy public life. In short, the principle of imitation and adaptation was something that helped many of these people in the cold countries to lead ahead in development. There is virtually no country in the world that had successfully developed with another's language; all have made it with the employment of its own language. Based on scientific research and well-defined goals, development can be done. It may sound interesting as the basic apparatus of language, the tools, may sour and knowledge from the talented are accessible to many societies. Development, therefore, is within the reach of *every society that is a bit serious*.

In conclusion, it can be asserted that the theory of adaptability has enabled us to substantiate that the origin of a superior species is a false notion. The theory offers us the knowledge of a preliminary state of human organism and how he evolved from his uncertain situations to a more orderly life and organisation. Human beings have one common denominator, and that is in terms of primary intelligence, the circumstances of the environment have much to do with that. Where there are specific environmental constraints, hazards, and difficulties that present a challenge to the organism, the mind through perseverance accommodates to find appropriate solutions to them. The use of original language of the organism quickened the manner by which he surmounted certain predicaments in his milieu, and consequently ensured easy imitation and adaptability.

Major and Minor Theorems of the Theory of Adaptability

We will now present regarding the theory of adaptability six theorems, which are of considerable importance in the acquisition of intelligence. All amount to asserting that intelligence acquisition rules out the false theory of a superior species; that the probable access to the primary intelligence and secondary intelligence should be preceded by the organisms development and use of his ***own blood language*** (i.e. the language that he or she is born with). This should be the language he should employ in the education of himself/herself. It also means examinations should be taken only with this particular language that is fixed in his blood and *flows naturally like water flowing down from a mountain top to the bottom.*
The six theorems are as follows:

Theorem 1. *If the brain develops in proportion to its encounter with harsh environment, and intelligence can be displaced where the environment is less demanding, it is the same as saying that the mind's development can be temporarily or permanently resisted where hardships are partially absent or obscured.*

This follows from Axioms 1 and 2. This is obvious from our argument concerning displacement. There was displacement that occurred in those areas where much time was spent in unnecessary repertoire; for example, building temples to house different gods, their sons and daughters. There were merrymaking and love making including sexual orgies in these temples. Where there was abundant food growing naturally nobody cared for finding ways and means to grow abundantly and save toward the future. No wonder areas in the tropics have been constantly affected by famine, since such hardships in terms of how to acquire certain basic resources were absent. But these were the cases in the colder climates before the use of science methods. When famine occurs once in ten years, no nation would ever take this problem seriously. And so nations that lived in the belt of abundance never thought in the first place to make serious plans about saving for the future. There were hardships or significant constraints, but these were obscured (i.e., because they did not make themselves directly available to human senses).
Q.E.D.

Corollary 1. Where there is displacement of intelligence, I say that there can be relaxation and this will make the inhabitants unable to plan ahead for their future.

Corollary 2. Hence displacement is associated with a decrease in productive activities that leads to high advancement.

Corollary 3. If your intelligence is displaced, it means that you are not serious.
Corollary 4. Hard work is not associated with intelligence displacement.

Theorem 2. *If chemical reactions can alter the state of organic bodies, then X-rays and radiation that possess inherent chemical agents can change/influence the genetic constitution of those individuals living in the extremely hot climates where the sun constantly shines.*

This follows at once from Axiom 2. This is obvious from numerous experiments that had been conducted since the inception of chemistry as a separate science discipline. The utilisation of X-rays in the laboratory for medical purposes also confirms that both X-rays and radiation have caused certain changes in the human organs. From the works of Boyle, Priestly, Lavoisier, Kelvin, Crookes, Roentgen, and Dalton, we have been enlightened on the effects of the new radiation in the laboratories. When X-rays, for example, emerged, it violated the paradigm of radiation theory with cathode ray tubes. The former was discovered only through somethings first going wrong with normal research. The change of the constitution is balanced by the allele that constantly produces pigmentation in order for the human organism to adapt to these destructive reactions that constantly take place in the hot climates of the earth (at least 5 hours sunshine daily).

Corollary 1. If these chemical reactions are indeed the cause of the physical as well as certain internal changes in man, then the organism had no part to play in these kinds of changes or influences.

Corollary 2. Hence these internal chemical alterations in man are caused by X-ray and radiation, and this is equivalent to certain reactions that human beings purposely create in the laboratories.

Theorem 3. *Given what has been proved, I say that if the individual who had probable access to primary intelligence and had further acquired certain characteristics from his utilisation of the secondary intelligence that aided in his imitation and adaptation, this will not make his successful adaptable qualities be attributable to his belonging to a superior species.*

This follows from Axioms 1, 4, and 5. Naturally there are important ingredients that account for the overall intelligence of individuals; this is totally in the domain of psychology. But this deduction of laws from axioms has one basic purpose: to ascertain what *fundamentally* provided intelligence for the primitive man in the inception of man's intelligence and development. These axioms above aid us to conclude with much strength that certain basic laws of the physical world had the influence in bringing the primitive man to utilise his brain, thereby accelerating the employment of the mind in a productive manner. We do, however, ascertain from these deductions that there is nothing to prove that some human beings were able to evolve developmentally because they were superior species. This is an invention, which is more an illusion than reality. This supposedly explains the reasons why there are always sabotages, conspiracies, biases, secret top meetings, and physically deliberate attempts to push themselves and their seeds forward in order to help in the domination which history tells us is temporary.

Corollary 1. Therefore, when the organisms through constant difficulties, crises, and frustrations succeed in the development of his environment, it makes adaptation easier and it elevates the organisms to a higher standard of living.

Corollary 2. Imitation and adaptation to the environment are ongoing process, and they will continue as long as the organism lives.

Theorem 4. *If intelligence is influenced by the demanding conditions of the environment and knowledge and use of science methods, and additional aid by the use of the original language (native) of the species, then IQ testing of two people from different species is only valid if and only if these two different individuals have each utilised their respective languages in educating themselves.*

This follows from Axioms 1, 4, 6, Theorem 3, and the **Secondary Law of Intelligence**. As we have already shown in the major discussion above, the use of the blood language contributed to the organism's acquisition of knowledge that ensured originality and some perfection in the carrying of some responsibilities. Unless an individual has been born and bred in another language that may be equally considered as fixed in him/her in a way, he/she may be handicapped supposing the language for learning is different. By these axioms that we deduce our theorem from we are capable of suggesting that intelligence tests should only do justice when it is applied to people that have only used their respective blood languages to educate themselves. In other words, let Candidate A with L-language as his blood language be compared with Candidate B with T-language as his blood language in IQ testing. Then we shall recognise this comparison as valid, since both candidates have originally educated themselves or gained the most part of their education through the use of their respective languages. Alternatively, let Candidate B utilise his T-language which is his own blood language and another Candidate C with N-language which is neither his blood language nor having been born and bred in it. We shall regard this latter kind of IQ testing or comparison as invalid and therefore injustice is being done to the Candidate C (the latter candidate). Originality and the sharpness we see in certain species that wrongly consider themselves as superior species stems from the fact that their blood languages, that are fixed in the blood, give them a great deal of advantage in imitation and acquisition of knowledge. When Latin was being used as the language of instruction and education as a whole, people were very ignorant and handicapped in those centuries. The return of Europeans to use their own various blood languages (i.e., German, English, Swedish, etc.) marked the beginning of the reformation and its subsequent great inventions and discoveries.
Q.E.D.

Corollary 1. If individuals had used their respective blood language in educating themselves, when examined on a particular test, differences will be marginal. Then justice could be said to have taken place.

Corollary 2. Hence the use of the language that is fixed in the blood of the individual increases his originality and sharpness in performing different tasks.

Corollary 3. Sharpness and originality in carrying out major responsibilities are associated with the use of blood language that we see in many species that falsely called themselves superior species.

Theorem 5. *With the same things being supposed as Theorem 3 and Axiom 4, I say that in most cases superiority is attained but not given.*

For let the individual be exposed to the primary laws as well as the secondary laws of intelligence in his community. Or, let nations follow the principles that are set out regarding imitation and the use of the primary, secondary, and main *frontis* guidelines, they will emerge as nations to be reckoned with in no time. And by this we mean to say that all individuals that become a bit serious in life and through perseverance organise their lives well, they will eventually succeed in this world. For no man has ever attained superiority by birth, all have been acquired through hard work and industry.
Q.E.D.

Corollary 1. Hence through perseverance and proper organisation superiority can be obtained by everyone/nation in this world.

Corollary 2. The principles of imitation and adaptation are the normal approaches that can help nations that are serious to attain development that leads superiority.

Corollary 3. If a nation has not been able to reach the height of superiority, then it is probable to view this particular nation that it has not utilised the principles outlined above (i.e., imitation and adaptation).

Corollary 4. All nations that have attained superiority have persevered and acquired it through hard labour.

Theorem 6. *Given what has been proved above, I say that development is the consequent of tremendous efforts by the whole population of a given nation, but its main impelling force is a few talented individuals that champion and inject wisdom into its acceleration.*

This is sufficiently clear from demonstrations presented above in connection with the primary and secondary laws of intelligence.

Corollary 1. This theorem holds for nations that have reverted into the use of their own blood language and as a result they possess the advantages that accrue from using natural talents.

Corollary 2. The inability to utilise ones own blood language can become forces of repulsion against proper talent development.

Miscellaneous

These theorems below will be proved or demonstrated at a later date.

Theorem 7

If obscure hardships could lead to intelligence displacement, then extreme hardships and stressful environment could consequently make the mind much more alert.

Theorem 8

If the uses of talents are involved in every nation rise to superiority, then all nations in this physical world have access to becoming superior in their development.

Theorem 9

If offspring ensure the continuity of a successful development, then every nation have access to continuity and successful development since offspring are available that are capable to taking over from their predecessors.

Theorem 10

If wisdom is found among every culture and its citizens, then Homo sapiens refers to the entire human race in its entirety.

Theorem 11

If talented individuals are required for every development in a progressive society, then every nation has access to these talented individuals that must be cherished and protected in order to guarantee their contribution to society building.

Theorem 12

If hard work is needed before any nation can develop, then those nations that are less serious will not be capable to developing to superiority.

Theorem 13

Intelligence levels between individuals can be bridged tremendously given that each individual subject is educated from birth with his/her own blood language (mother tongue).

Theorem 14

If an individual has mastered his own language very well, the amount of time taken to master other languages that is not his own mother tongue is shorter than otherwise.

Theorem 15

Given that all individual subjects utilise their own language in the acquisition of knowledge, this will put all subjects in equal footing and there will not be any major difference in intelligence between these individuals.

Theorem 16

The intelligence of two people that have acquired education with their mother tongue may experience unbalanced nature if, say, one had poor training that affected his language ability and comprehension.

Theorem 17

If those who have not been serious are lacking behind in development, then those who have managed to develop successfully are the most serious ones.

Theorem 18

If those who are less serious have not developed to superiority, then those who are serious are the ones who have certainly developed to superiority.

Theorem 19

If the use of blood language can guarantee a successful imitation and adaptation leading to quick development, then the employment of the blood language is an ingredient of being superior.

Theorem 20

If not being capable to using the blood language in imbibing knowledge could lead to being handicapped, then access to one's blood language in learning could prevent handicap in these areas.

Theorem 21

If chemical reaction could alter the organic states of substances, then X-ray and radiation could change the appearances of matter in the extreme hot climates of the world.

Theorem 22

Man, the great seeker, and his achievements, whether for food or for proper adaptation, have been without exception the product of his chief distinctive quality, his power of thought.

Theorem 23

The nuclear reactions that goes on currently in space, and within stars like the sun transform hydrogen into helium and their effects are felt on the planet earth.

Theorem 24

Hydrogen is one of the basic elements that matter man is made up of, and it consists of a heavy (electrically positive) proton and a light (electrically negative) electron. This element is susceptible to the burning heat of the sun.

Theorem 25

That the earth is in fact irradiated from all sides by radio waves with the anticipated properties lends support to the picture of an early very hot, phase of the universe.

Theorem 26

The analysis of the evolution of stars that have changed most of their hydrogen into helium shows that they will become dense and not enough for more complicated nuclear reactions to occur, like those which change helium into carbon. This is reminiscent of how chemical reactions can do to alter the states of matter and other substances.

Theorem 27

Superiority in development is not attained overnight, in fact, it is a gradual process that requires careful planning through the help of talented individuals; and this does not make any successful nation to be regarded as made up of superior species.

Theorem 28

To make a valid test of students on intelligent tests (IQ), it is required that all candidates must have utilised their blood language in imbibing knowledge, which put them on equal footing.

Theorem 29

All nations that have developed to superiority have laboured hard and had achieved it through difficulties, frustrations, hopelessness, and much help was also received from other nations abroad.

Theorem 30

The view that a particular race is superior and can dominate all others is a false notion, and can be cherished or believed only by deluded individuals whose infantile beliefs make them constantly insecure due to the threat they face from without.

Box 7 Formulae for successful development leading to adaptability

---

De = development
Im = imitation
Ct = cultural context
Ad = adaptation

$$De = \frac{Im + Ad}{Ct}$$

Dp = displacement
Pf = primary frontis
Sf = secondary frontis
Mf = main frontis
St = strictus
Steo = science theory
Smt = science methods
BL = language
Int = intelligence
Pθ = primary intelligence
Sθ = secondary intelligence

$Dp = - (P\theta Int + S\theta Int)$; Formula for intelligence displacement

$St + P\theta Int$; Formula for primary law of intelligence

$Steo + smt + BL = S\theta Int$; Formula for secondary law of intelligence

$St (P\theta + S\theta) + BL = Int$; Formula for intelligence

$\int Int = st (P\theta + S\theta) + BL \int De = \frac{Im + Ad}{Ct} \int Pf + Sf + Mf$

Formula for successful development

Scholium

All the important laws of the theory of Adaptability have been deduced from the seven axioms, which comprise the Primary and Secondary laws of intelligence. These six theorems, which have been rightly demonstrated, enable us to enumerate some of the *laws of nature* that have influenced mankind to survive and adapt to its environment. The diligent reader will recognise that the theorems are necessary consequences of the axioms, so that if the axioms are accepted, the theorems must certainly be accepted also. The theory has provided satisfactory explanations to some of the vital baffling questions which cause problems in this modern world. Surely the theory will raise further interesting problems that will call the attention of scientists, both great and small; to earnestly pursue research concerning the laws deduced from the adaptability theory. Let this theory be put to excellent use in this modern technological world. And that without entertaining any partiality, all researchers who happen to examine this theory will lend their ears and also test it. Let it be examined critically to see how its comprehension will enable us to stop injuries to other people who have not reached the stage we are at. This is imperative when the theory's predictive power should allow us to accept it that superiority is only attained through hard work and development, and *has nothing to do with innate ability*. Like education and the organisms' development, a little seriousness will make every society that is aggressive and earnest, as well as persistence, attain superiority, as it is happening in our very face in the Far East, and more specifically regarding the sleeping Giant, China. Test this theory for its truth or falsity, and accord it a place in the next millennium as a promising theory to be accepted or worked upon.

Chapter 9: Of Systems of Political Discriminology

Political discriminology is considered part of the science of discriminology that equips the statesman or legislator with potent knowledge on how to provide for the overall subsistence, both economic aid and employment, to the *subjectuses*. It is a science that makes its utmost objective to caress the *subjectuses*, which is in Latin *carus*, meaning 'dear.' Political discriminology proposes to bridge the gap between the *subjectuses* and the native citizens in order to enrich both the people and sovereign.

Of the Principle of Restraints

There is only one world that we know is habitable at the moment, even though there are several worlds that our expanding knowledge has given us the opportunity to detect. But until these worlds are properly prepared for habitation in the future, there remains only one world, our dear earth, to live and cherish. The movement of people across borders to seek pleasure in other climatic regions in certain periods of the years tells us that human beings, unlike animals put in the cage at the Zoo to be watched, need to move freely because we love freedom. Even birds in their free world in the air give us the idea that there is joy in flying or migrating to anywhere they want to, so far as the weather is fine and warm.

While the regulation of commerce can increase the quantity of industry in any society beyond what its capital can maintain, the regulation of human beings when carried to the extreme infringes the freedom of the human organism. Too much regulation can divert the interests of people into the direction which it might otherwise have not gone. That is, it may affect the economy and the make up of the population that is likely to be more disadvantageous to the society than that which it would have gone of its own accord. Every individual is continually exerting himself/herself to find out the most enriching place on earth whereby it will be advantageous to live in and work or raise up his children. In the past the nomadic people, who investigated such areas that are fertile and good for pasture for their cattle, practised this. They believed the earth to be no man's land so that they could move around graciously while still preparing for the support of their family and their nature-given world. There were no restrictions whatsoever, except occasional wars that may be fought with resistance, and even this, it was not because they moved around or migrated freely, but because their possessions had attracted those lazy idle individuals who would prefer stealing than working to acquire something for themselves.

Of allowance of the *subjectuses*

Thus upon equal liberty and equal right should we consider all the people that have migrated into another land legally with their permission granted the right to work and use their natural given talents. There are many people that prefer to engage in trade abroad and enjoy even having the consumption of goods that come from abroad than the ones produced in the home market. If we have no problem in allowing common goods produced abroad in entering into the country, how much more should we have some restraints regarding the flow of human beings into a particular region of the earth. From the consumption of goods one can also say something about the "consumption of human beings," that is the liking for other human beings. It is not always trade that many people want to engage in outside their country or abroad, but there are many people who, without the importation of people into its country, would not have been successful in acquiring friends, husbands, wives, adopted children, and also workers. Our knowledge of these things and the commonness of these contacts make us forget that there is some richness involved in allowing human organisms to be part of other groups already living in the land. Before nationalisation, people had the greater opportunity to move where they wanted and live where they wanted. It was great, memorable periods, where human beings intermarried and God knows our different species could produced, variations that made and resulted in diversity of inhabitants of the world. Our Kings and Queens had their origin in far distant lands if we were to trace their histories, and also open up the great mysteries where our great-great-grandfathers or mothers came from in the beginning. The knowledge of needing one another is still cherished by our parents, those that take time to read and acquaint themselves with the history of our great descendants. The idea of looking for a better breed to join certain countries is *madness* that is cherished by only those who are ignorant of the past.

Of unfairness of restraint measures

The idea of putting restraints on the intake of certain *subjectuses* into other counties is always criticised by some organisations that have made it their responsibility to do this check. This follows the supply and demands principle, where need of certain professionals is said to determine who should migrate to another part of the world. This is unfair, as usually those that migrate have a better chances of surviving in the country they originally resided. In such rationalisation one country looses, the other gains. In other words, the idea of accepting only those with a profession benefits the host nations rather than the nations that the *subjectus* is leaving behind. These unfair restraints are for those poor nations who always train their citizens with meagre resources and then loose them in the end for the richer nations. This trend is also seen with certain richer nations who train their *subjectuses* and later loose them to their neighbouring nations that see more use of the resources of these people. In the latter, there is the self-sufficiency principle that makes these nations reject a large number of people who could have injected strength into their economy. These restraints, though favourable to the thinking of the majority of the natives residing in these nations, the long term effect is detrimental to the economy, probably more damaging to a well-planned economy than other market economies.

On the question of revenue and the *subjectuses*

Nothing is more usual among societies that possess superiority complex that have made some advances in commerce, than to observe on the progress of its *subjectuses* with a suspicious eye, to think all its foreign organisms as their rivals, and to assume that it is impossible for any of these fellows to flourish, but at their expense. On the contrary to this narrow and malignant notion, I will venture to assert that the increase of riches and even properties of all the *subjectuses*, instead of hurting, commonly promotes the riches and commerce of a particular society's citizens. And that a state can scarcely pursue its trade and industry very far, where all the *subjectuses* in its society are buried in poverty, ignorance, and disillusionment.

It is correct that the economy and industry of a people cannot be disturbed by the greatest prosperity of its *subjectuses*; and as this branch of commerce is undoubtedly the most vital in any country, we are so far removed from all reason of jealousy. It can be further observed that where an open dialogue is conserved between a nation and its *subjectuses*, it is impossible but the interests of everyone receive an increase in focus as well as from the improvement of every one in the society.

Compare the situation of the USA at present with what it was some centuries ago. In the field of both arts and sciences it was not advanced; in fact, it was lacking in perfection. The British scorned their system of education as inferior, if not poor. Every improvement they have made since the depression has come from the influx of foreigners from chiefly Scandinavia, Asia, Africa, and Europe. This interaction had been a great advantage. From the University Professors, Scientists, Doctors, and nurses, famous researchers like Albert Einstein, Hannes Alfvén and Von Braun who invented the rocket that took America to the moon, America had tremendously advanced in many things. All because America is not afraid to employ the services of those they think have something that can contribute to nation building. They do not panic because too many academicians are becoming politicians, or do not ask where someone who is migrating to their country hails from originally. The *subjectuses* in those countries that accept them whole-heatedly as one of them continue to excel, contribute to their industry, invent new machines; in short, they hold the country in the uppermost high. Notwithstanding its various politicians with diverse cultural backgrounds, who perform wonders to the astonishment of the whole world.

The increase of *subjectuses* lays the foundation of foreign contacts and commerce. Where a great number of these subjects are permitted to be in high places it broadens the contacts of the host country, as it will have diverse markets for their products. These could be exported with advantage. But if the *subjectuses* have no part to play in politics and do not find positions in the governments, they will have nothing to give in exchange. The *subjectuses* are in the same position as machines. When they are not used regularly they wither and begin to rust. A person that thinks himself as industrious becomes idle where he sees his potentials being neglected day in and day out in the job market. The riches of several people in the community contribute to augment the riches of the individual members, whatever profession the latter follow.

But what if a nation should say we do not have jobs in the market, such as sometimes happens in many countries with well planned-economy, must not the employment of the *subjectuses* causes problems for the citizens? I answer that, when there are no jobs in the market, it is supposed to be a natural advantage for the ruling government to commence finding alternatives to aid to raise the economy. The problem of scarce jobs should not make the legislator idle or any government to blame its *subjectuses*. It simple means that there should be increase in industry at home with respect to other manufacturing products. The consequence will be increase in activity in industrial sectors and a renewed spirit in the area of manufacturing. We should never be afraid to have our *subjectuses* remain in equal footing with our citizens. The emulation among subjects of one nation serves rather to keep the spirit of work morale up; it keeps the industry alive in all times. And any people are happier who possess subjects from diverse backgrounds, than if they remain homogenous or come from a single race.

Nor needs any society entertain the fears that their *subjectuses* will amass wealth such that it will be very difficult to control them and have them obey the laws of the land. By giving a diversity of geniuses, climates, and soil to nations, nature has given us the freedom to move around to wherever we can offer our services in return for a better life. The skills of the *subjectuses,* if properly tapped could make any nation gain tremendously more than they the *subjectuses* could win from the state. And so the subjectuses should be *encouraged, caressed,* yea *wooed,* to stay behind, instead of jubilation when we see them pack their luggage and leave the country. Only ***improper politics*** and ***ignorance*** will make us be in jubilation when we discover people with high education, instead of being absorbed in the job market, leave for our neighbouring countries, where they find them useful. It is not a matter of racism, but lack of foresight of those people in the responsible positions who fail to do their job. By doing this they disappoint us electors and we, the taxpayers, who have given them our confidence. *I challenge that where we see many natives, yea more subjectuses, leave the country in question, it should be interpreted as lack of responsibility on behalf of the politicians instead of jubilating that "we have a way of sending all of them out."* The politician that has been put in that responsible position should be reprimanded.

Were our narrow and malignant politics to meet success, we should reduce all our *subjectuses* to the same condition of sloth and ignorance that prevails in most underdeveloped nations in the world. But what would be the consequence? They could send the common safety principle into confusion. I wish that most nations would recognise the need to let its *subjectuses* flourish just as it is happening in some pioneering countries in the world today. They will dominate, as these do without much difficulty.

The Legislator and Statesman

As has already been mentioned, the *subjectuses,* like all other subordinates of the sovereign, need the tranquillity of mind which originates from opinion that each member of the community also possesses his safety. This is the source of political liberty. It is for this liberty that the legislator and statesman should work toward, so as to ensure that the *subjectuses'* safety is guaranteed in a manner that the latter would feel safe to enjoy his liberty.

The laws of every country are necessary ingredients of its democracy, and these are important for the *subjectuses* as well. The methods of communication of these laws to the *subjectuses* are important so as to enable ignorance to disappear from among them. There are many individuals that have made a quick shift from a land where so little regard is given to the laws; these individuals should be aided to know the importance of these laws. Where the *subjectuses* have migrated from an Islamic state, the legislator should ensure that these people are instructed regarding the laws, as for this there is a proverb that says that "If you travel to Rome, do as the Romans do." For the danger exists that by neglecting the instruction of these laws to the *subjectuses*; he may for sheer ignorance cause injury to himself and other citizens. There is, therefore, even here the necessity that the legislator that is more concerned with the situation of the *subjectuses* be more conversant with the tyrannical laws and systems of governing of these countries where the different people that made up the *subjectuses* hail from originally. Otherwise where violence and oppression are the system of governing this may be transplanted in the host country in question.

Criminality and discrimination have affinity with each other because one is perceived to cause the other to emerge. In fact, *if you cause discrimination you create criminality among those in the marginalised segment in the society. Conversely, if crimes and misdemeanour are allowed to prevail in any society, the middle class and the upper class citizens will hate those in the grassroots all the more, and the consequences will be discrimination or segregation among a large section of the population.* Many individuals that found themselves breaking the laws have some conditions that relate to discrimination. The background factors reveal that individuals have been affected because their parents have constantly been discriminated against in work places and other areas of the community. This cycle may continue that, when due to their loosing their jobs the children are compelled to survive on their parents' meagre income alone. The legislator or the statesman must confront such hidden factors that originate the problem of criminality seriously. It should be their utmost prime business that they fix their attention on these factors that originate criminality, and also discrimination in the community. The security of the *subjectuses* should be seen in its wholeness perspective, and be taken up in order to guarantee real proper liberty for the *subjectuses*. But should the legislator think itself as not interested in these matters, the danger is some conspiracy to let escalate into a serious problem by the underground groups that will threaten the safety of all peace-loving people.

Scholium

We have seen the danger that inequalities that exist in societies' citizens could bring to any democratic government. Our knowledge of these facts should make us all the more bent toward a better integration that would guarantee the interest of all, and also make us bridge the gap between citizens of a country. Our prejudices should not make us consider only natives in the country as needing this help to maintain appropriate equilibrium, but we should strive to maintain the desire that enable us to put severity and unequal treatments aside or behind us and look for the betterment of all.

In what has been asserted so far the legislator or statesman should make his priority aim to lend his ear to complaints of the *subjectuses* and to tackle the problems that make them go astray from observing the laws of the society. Even if we have not mentioned all that are supposed to be the inner yearnings of the *subjectuses,* we have succeeded in the foregoing to have outlined the essential needs that any knowledgeable society should make a priority to provide for its *subjectuses*. There is a need to investigate their situations continuously in order to unveil them through research and consequently give currency to the problems that are hereby to be enumerated.

First of all, the principles we have proposed here are not new, but yet vital for the statesman or legislator that has immersed himself/herself in political discriminology. That if it is necessary and more appropriate:

1. To provide better economic aid to the *subjectuses*;
2. To promote a well and measurable subsistence;
3. To bridge the gap between natives and *subjectuses*;
4. To see that the *subjectuses* are a resource to the nation in question;
5. To expand their effort into the labour market with the purpose to absorb them;
6. To give appropriate education so that they do not become useless in the system;
7. Better training for them to behave and think like the authorities want them to do, so that they can "fit" inside the society:
8. There should be a tangible approach and useful way to convey them more closely to the society in order to make them more useful to the people and the sovereign;
9. There should be a genuine search and constant regulations to put them to a better use in directions where each individual has already propagated himself or herself;
10. To promote public interest among them so that it can generate a patriotic spirit among them;
11. The constant desire of the government to generate annual revenue from them that the government's expenditure does not exceed the revenue accruing from their own effort;
12. That they should be made conscious that they can contribute to the capital of the government/society;
13. To assimilate requires constant attempts to get them involved and make them stand on their own feet;
14. This means that unnecessary restrictions put on them that make it impossible for them to obtain jobs should be lifted so that they will reach their goal;

15. This therefore requires that appropriate planning and care be taken so that no underground or interest groups can sideline the government's' effort to make the *subjectuses* feel that they are vital to the economy and are at the same time being cherished.

These points mentioned above should be rehearsed by all that will make it their objective to care for the *subjectuses* in their society. It also sounds the need to get prepared before any society makes the decision to accept more of these necessary individuals into their midst.

General Scholium

In every branch of science, now and then important discoveries are made that make scientists gain extraordinary insight as to how its continuous development should be directed. Of late that negative human physical behaviour called discrimination has received impetus that makes its scientific investigation imperative. Discriminology is the branch of modern science that contemporary scholars contend would gain the attention of young scholars in physics, astronomy, and mathematics that are interested to investigate the "matter man." Especially now that the world is moving in the right direction whereby the right of the individual as well as societies are continuously encouraged to be respected.

Science is a unique aspect of man's development, as this had generated constructive thinking that had altered the physical world, especially during the last century. The scientific method, I say, is the most assured technique man has yet devised for controlling the flux of things and establishing stable beliefs. The method of science does not seek to impose the desires and hopes of men upon the flux of things in a capricious way. It may indeed be utilised to satisfy the desires of men. But its prosperous employment depends upon seeking, in a deliberate way, and irrespective of what individuals' wish are, to detect, as well as to seize the advantage of, the structure which the flux possesses. To apply the scientific methods to investigate how to live with one another in this planet is a unique step forward. It provides the most important opportunity to question certain long-held beliefs that have been accepted by people so many years ago without critical examination.

Scientific method pursues the road to systematic doubt. It does not certainly doubt all things, for this is clearly impossible. But it does question whatever lacks sufficient evidence in its support. Moreover, science is not content with psychological certitude, for the mere intensity with which a belief is held is no guarantee of its truth. Science demands and looks for logically sufficient grounds for the propositions it puts forward.

In our treatise, sensitive themes have been touched upon when we set out to develop these vital theories concerning discrimination. These theories are considered to be significant to the entire history of mankind and in the attempt to fight for man's survival on the planet earth. The ability to formulate problems whose solution may also aid solve other problems is an unusual gift, requiring extraordinary genius. The problems, which meet us in daily life, can be solved, if they can be solved at all, by the application of the scientific method. But such predicaments do not, as a rule, erect far-reaching issues. The most striking employment of scientific method is to be discovered in the various natural and social sciences.

When theories are developed for the first time, not everybody would accept them as authentic or were capable of explaining everything that they are purposed to elucidate. But later when scholars put them to test and discover them to be useful, people begin to accept them and then accord them a unique place in man's history. It is then that the theorist may be enabled to enjoy his status as a proponent of a useful theory that demonstrates what it set out to explain. Let us bear in mind that these theories are in their preliminary development stage and that much work is needed before they will become perfect in this world filled with copious doubt. The enterprise of science is a self-corrective process. It appeals to no special revelation or authority whose deliverances are affirmative and ultimate. It claims no infallibility, but relies upon the methods of developing and testing hypotheses for assured conclusions. The canons of inquiry are themselves detected in the process of reflection, and may themselves become modified in the course of study. The method makes possible the noting and correction of errors by continued application of itself.

It is usually the case that some time is required for theories to be examined and to be accorded their place in the scholarly world. As I pen these words I am not at all surprised that it would take some years or probably centuries before we will see the use of these modern scientific theories. So, as a matter of fact I am not discouraged. All pioneers who contributed to such unique endeavours never had it easy. Some never lived to see their ideas put to use in this physical world. Others only began to perceive their theories' importance when they were on their deathbeds. But others were lucky enough as they lived to see their theory become accepted, and even had considerable time spent on its development to make it much better or workable. Let us hope that these theories that signify a big farewell to discrimination will gain attention from scholars, who not only will criticise, but also will offer contributions that will aid these theories to be well accepted. It is the firm hope of the proponent that these theories will be sufficient to break down the great walls of discrimination in this modern world.

Mathematics is required to be utilised to find those quantity of forces and their proportions that engineer the conditions that result in this negative behaviour during war and peacetime. Then, coming to physics and chemistry, these proportions must be compared with other phenomena, especially those that apply to the reactions that affect the entire existence of mankind. And then, finally, it will be possible to utilise astronomical findings to deepen man's understanding of himself and his relation to other created bodies of the universe. Let us see, therefore, how the suggestions to utilise these disciplines that apply rigorous methods to comprehend the physical universe be transferred to the investigation of the matter man in connection with his negative physical behaviour of discrimination.

APPENDIX A

Perspectives in Relation to Learning and the Theory of Blood Language

Preamble

Once again we come to the old strong contentions that exist among scholars concerning how learners proceed complexly to acquire or imbibe their own learning. It is a characteristic of scholars that we do not let sleeping dogs rest; instead, now and then a long-standing problem is revisited to detect whether we can illuminate a brighter light on it that might also contribute to correcting our suspicions or to put the mind in equilibrium. Furthermore, experience has educated us on the vitality of constantly giving prudential considerations to such matters of scholarly interest that affect our whole fabric of existence and human beings technological development.

    This is all the more reason why I am resolved to renew a fresh interest concerning how knowledge, human beings' greatest weapon, is gained. Now this manner in which students secure their learning in the classroom has been the focus of major discussions among researchers and educators alike for some time. In the midst of their intensive discussions, it appears students' thinking in the classroom had been prominent as the object of research. In the reports issued in connection with classroom learning, students' thinking had influenced earlier classroom research, and the categories and concepts utilised by these researches had been borrowed from the work undertaken in the psychological laboratories. Research embarked in the laboratory was geared towards the setting of distinct boundaries between different types of behaviour found in the classroom. Nowadays, distinctions between thinking, learning, and achievement are no longer of special interest, as it came to its zenith during the latter part of the twentieth century.

    According to Nuthall's influential article "Understanding Student Thinking and Learning in Classrooms" published in (1997), *International Handbook of Teachers and Teaching,* classroom research that appraised and studied learning, employing psychological categories and concepts to demarcate the boundaries of diverse types of classroom behaviour, regarded thinking as higher than learning. Because of this, "thinking" that is conceived as a "distinct set of processes" that functioned in the brain of the learner and mediated between instruction or teaching and behaviour in the classroom and learning, received tremendous attention.

    It seems the current alteration in concepts of thinking and learning in the classroom has been caused by new perspectives that have been introduced from a range of diverse disciplines such as social anthropology, linguistics, sociolinguistics, semiotics, and many others. Different methodologies have also been utilised such as textual criticism, phenomenology, phenomenography, and ethnomethodology. New outlook regarding conceptualising and comprehending the experience of the learner in the classroom has also emerged from psychological studies undertaken in the former USSR.[69]

The Purpose

The objective of this chapter is to make investigation into the different perspectives of learning and to unveil in what manner each of them contributes to our knowledge and understanding of learning and classroom communication. *Attached to this purpose is shortly to present the Theory of Blood Language, which I propose as a necessary factor or theory in the acquisition of learning.* At issue are these questions: What do these different perspectives inform us about classroom learning? Are new perspectives of learning currently being developed? What are the central concepts of these perspectives on learning? What are their similarities or differences? Are there some major points of conflicts between these theoretical perspectives? What criticisms do scholars of different approaches have for one another? Finally, how can the new perspective of the Theory of Blood Language assist us to comprehend how individuals can acquire learning in the social context?

Different Perspectives on Learning

Numerous studies of students' experience in the classrooms have enabled us to successfully unveil the prevalent themes and issues that bring these perspectives together. In what follows, I shall deal with each perspective separately; that is, analyse the individual perspectives in the attempt to reach new ideas connected to them. Later I would proceed to commence a general discussion concerning how a blending is necessary to provide us a much better perspective on classroom learning. In this case, then, my objective is not different from many articles presented earlier on this topic in other journals; but as indicated already, not ignoring such approach to study means that we are not yet satisfied with the existing explanations. Or we may augment our confidence in them through their subjection to this meticulous analysis.

The first is the *cognitive constructivist perspective*, which entails studies that utilise psychological concepts in their approaches to the study of learning. Here, learning and thinking are fused into the cognition concept. Students learning are seen as resulting from the way the students themselves create or construct their own knowledge and skills in the classroom. The second category is the *sociocultural and the community-focused perspective*. This category is said to encompass studies that are basically sociocultural in their approach. In other words, this particular perspective sees learning and thinking as social processes which take place in social contexts between individuals. Learners progress through a process of "apprenticeship" within significant social groups. The third perspective is named as the *language focused perspective*. This deals with those studies that have language or sociolinguistic in their orientation. Language in the classroom is seen as entailing both the content and the medium of learning and thinking. The gains made by the students in classroom study are the linguistic "genres" of the disciplines. Enshrined in these genres are the concepts and manners of perceiving and thinking that are characteristic of the disciplines. The fourth perspective is *phenomenography and variations perspective.* This latter perspective is more recent. Its advocates come from the University of Gothenburg, Sweden. This is an orientation which identifies, formulates, and tackles certain kinds of research

---

[69]Nuthall (1997) sees the work of these former Soviets scholars as have injected new thinking into knowledge acquisition in not only psychology but also a whole range of disciplines.

questions related to learning and understanding in an educational setting.[70] Phenomenography purposes to identify and describe the various ways in which people experience, understand, or comprehend certain phenomena in the world. In the phenomenographic orientation, learning is seen as a change in the learner's capability of experiencing something in the world. In other words, learning is a change in the structure of the individual's awareness.

The Cognitive Constructivist Approach

Jean Piaget and his followers

Adherents to this perspective subscribe to cognitive science and the concept of "cognition" and state that students create or construct their own knowledge and skills. This perspective is built upon the theory of a French philosopher and a biologist, Jean Piaget. The work of Piaget and Inhelder[71] discusses in detail Piaget's theory, and illuminates on the cognitive structures, the mental processes that the developing child possesses which aid it to learn. Piaget means that a child undergoes certain different stages until it matures to a final stage where it attains some kind of cognitive maturity. Piaget seems to confirm the idea that the child at the period of Sensorimotor, that is, the first stage, shows a certain kind of intelligence. Activities such as purposeful groping, sudden comprehension or insight, co-ordination of means and ends, and so on, attest to existence of intelligence before language. This intelligence empowers the child to solve problems of action; for example, reaching distant or hidden objects by constructing a complex system of action-schemes and organising reality in terms of spatio-temporal and causal structures.[72] As the child lacks language or symbolic function, these constructions are made with the sole support of perceptions and movements, and also by means of a sensorimotor co-ordination of actions, without the invention of representation or thought. As can be seen, Piaget's studies indicate that there are some fundamental forms of knowledge with the child and these are named as "assimilation" and "accommodation". Assimilation means that "reality data are treated or modified in such a way as to become incorporated into the structure of the subject." This "filtering" or "modification" of information is sent to the internal schemes to contain reality and this is what Piaget calls accommodation. It is this idea of the existence of intelligence before language that I shall return to in order to present Vygotsky's criticism of Piaget.[73]

We see here that the cognitive constructivist perspective has its starting point from Piaget's studies and his theory on cognitive structures. This view, that learners construct their own knowledge as they engage in the processes of interpreting and making sense of their experiences in the classroom, is therefore based on earlier observations made by Piaget of the child. Learning is regarded as the conceptual restructuring that results from this cognitive processing.

One question that scholars pose in connection with this perspective is: If learning is acquired through restructuring alone by the student, then what is the function of teaching? It is not surprising that the consequence of this view makes some doubt whether there is any direct connection between teaching and learning. For example, the manner that the tasks are arranged, the questions the teachers ask including the exercises the students practice; these can only have indirect effects on student learning. Nuthall, however, does well

---
[70]Marton & Booth (1997).
[71]Piaget and Inhelder (1969)
[72]*Ibid.*
[73]*Ibid.*, pp.16 ff.

when he explains that teaching, however, does its job in the students. "As students encounter new experiences, their minds construct representations of those experiences that are structured by their own previous knowledge and beliefs. These individually-constructed representations interact with each other in the production of new knowledge and beliefs."[74]

Problems: classroom learning in the constructivist perspective

Researchers who adhere to this view have tested this perspective in the classroom, learning and they have found it to be workable. The interview method has been employed to study the interaction in classroom experience between the teacher and his students. Students have been observed as indeed creating or constructing their own knowledge and skills. In the empirical studies that have been carried out by researchers in this field, three research questions have been noted to be the prime concern. In the first place, they ask themselves how classroom work and activities of practical nature can help structure cognitive processing. Secondly, how do the social processes of the classroom affect the constructing of knowledge and learning that takes place in peer interactions in small groups or corporate work? Thirdly, what role does the teacher play in motivating, structuring, and guiding the learner's cognition?

According to Nuthall, the three questions or variables have been the main concern of several studies in this area in the classroom. With regards to the first variable, that is, classroom tasks, research has unveiled that there are two major problems that relate to the content of tasks carried out by students in the classroom. The first problem is the transfer of knowledge to their understanding of everyday events. In Mathematics Education and Science, it has been observed that "effective cognitive engagement within the context of a specific classroom activity may not produce the understanding or knowledge needed in later learning or different contexts."[75] It is the authentic problem-solving in the classroom that is the only solution to this problem of transfer. It is concurred that the nature of the problem decides the sort of learning that takes place. The other problem is the significance of the content being studied.

Peer interactions have been explained to have a privileged role in changing student's cognitive structures. The reasons given are twofold: (1) the difference in perceived status between adult views and student views; and (2) the interrelated nature of the cognitive and the social in peer relationships in the classroom. According to Nuthall, one notes that cognitive processes and social relations are closely mingled when it comes to learning at school, that one cannot easily put a demarcation between them. Adherents of the socio-constructionist perspective, for example, Doise, Doise & Mugny,[76] Nuthall notes, utilise this view as the basis for a separate theory of classroom learning.

Finally, studies concerning the role of the teacher among the students' constructing of knowledge and skills have enlightened us on the role of the teacher as the motivator or facilitator of student thinking. The teacher is also said to be the source of established and effective knowledge. If we follow this perspective's description that says that students' learning is acquired through the "residue" of the mental processing that transpires in the classroom, then I also agree with those who see that there is a major obstacle that faces the modern teacher. The problem entails the teacher being able to structure and lead the learners' mental processes so that the "residue" of these mental processes will be consistent with desirable outcomes of the curriculum.

---

[74] Nuthall, 1997:684.
[75] *Ibid.*, p. 691. Hiebert *et al.*'s (1994).
[76] Cf. Doise (1991) and Doise & Mugny (1984).

So much has now been said about the cognitive constructivist approach and how they understand learning. What about the sociocultural approach; how do they also comprehend learning? The question that we should bear in mind is; in what way does the next perspective differ from the foregoing? What at all makes the sociocultural and interactionistic concepts unique?

Sociocultural and the Interactionistic Perspective

Those who favour the sociocultural perspective have alleged criticisms against the adherents of the cognitive constructivist theory. These critics assert that the latter is too closely tied to a "technocratic," "positivistic and individualist perception" of education and the social order.[77] What they claim to be the greatest weakness of the cognitive constructivists is that they have turned a blind eye to the graphic roles that the social relations, community and culture play in cognition and learning.[78] The sociocultural perspective's strongest advocates are Lave, Rogoff, Lemke, Wertsch and many others. Later on in the study I shall deliberate on the importance of the work of the Soviet psychologist Vygotsky, who contributed tremendously to the core concepts of the sociocultural perspective. His work discusses the social origins of intellect.

The foremost advocate of the sociocultural perspective is Lave. Lave, for one, does not comprehend why mind and rationality, while being the pivotal concepts of the cognitive constructivists, could be denied consideration from the activities and settings in which they might be identified. He contends that, among other things, cognitive contructivists are wrong in their approach to solely concentrate on learners as individuals who create their own knowledge and keep it as "disembodied mental representations in memory."[79] Mind, rationality, and knowledge are themselves cultural concepts. A call, therefore, is made to review the way we conceptualise socialisation of students. Lave's criticisms are based on the support provided by her studies conducted in both Western and non-western schools settings. Her proposal includes her suggestion that the acquisition of culturally significant knowledge or ideas should be recognised as the process of becoming a member of a community. Lave also believes that the learner does all in her power to acquire an identity as a member of a "community of practice." In her utilisation of the term, a "community of practice" entails a class of people who are known as possessing a special knowledge or competence in some branch of cultural practice (e.g., teachers, hockey players, nurses).

The perspective adopted by the sociocultural adherents, especially Lave, therefore defined learning as a process whereby identity is formed within "a culturally structured community."[80] This perspective also implies that the conception of mind and knowledge are processes that take place as human beings interact with each other in different activities organised in the community. Rogoff's definition is not different from his counterpart, Lave. According to Rogoff, learning is seen as a function of roles' transformation which takes place gradually as an individual takes part in the activities in the community thereby becoming an experienced member of a community of learners.

Perhaps the most important concept, which both Lave and Rogoff utilise in their analyses of the above perspective, is the concept of "apprenticeship." This concept indicates the active role the child subjects himself to in initiating and organising his own learning. It is

---

[77] Nuthall, 1997:700.
[78] Dewey was well aware of the social nature of learning. He stressed the need of adopting a social perspective on the whole educational enterprise (Phillips & Soltis, 1998). School was seen as a community but educators have overlooked this and instead often kept student isolated at their desks.
[79] Nuthall, 1997:700.
[80] *Ibid.*

the socially interactive manner of the learning process whereby an experienced individual see to it that the child gains knowledge through the process that is structured by technologies and cultural institutions.

The Contributions of Vygotsky

Vygotsky is a scholar whose name is associated with the sociocultural perspective of learning. First, we are made to understand that sociocultural perspective owes much to him because it does not become easy to comprehend the concepts of this perspective without taking his work into consideration.[81] His work examines the social origins of the intellect. Through him the nature of the human mind was explained, and this he did by tracing the mind's development and maturation in the child. Second, Vygotsgy is known for his strong criticism of Piaget's theory.[82] His criticism was on theoretical as well as on methodological aspects.

Vygotsky disagrees with Piaget, who thinks that the child's development occurs only on a psychological level. For Vygotsky, cultural development of the infant takes place on two levels; the first is the "social level," and the second occurring on the "psychological level." These developments reveal themselves between people as an inter-psychological category and later within the infant as an intrapsychological category. Observable developments that take place can be named as concept formation, logical memory, volition development, and voluntary attention.[83]

Vygotsky means by the "first level" that the origin of the individual mental processes has its roots from the social interactions that occur between the child and significant others. Vygotsky argues that social interaction provides the child with its forms and structures, which are not different from the forms and structures of higher mental processes.

Social learning was the young person's ability to learn by imitation. Interacting with adults and peers in co-operative social settings gave the young learner ample opportunity to observe, imitate, and subsequently develop higher mental functions.[84]

The psychological and technical tools in the culture most often determine the forms and structures. The tools the culture provides entail cultural artefacts such as works of art, maps, mathematical symbols, and technical machinery. But the most important of all tools

---

[81] A summary of Vygotsky contribution is as follows: Vygotsky contributed to the socio-cultural theory in the following:

- We learn from others those "psychological tools" that human societies have invented to allow individuals to deal effectively with each other and the world.
- Logic, symbolic transformation, concepts, forms of notation, signs, numbers, and words and also working tools like the hammers and saws of carpenters. These are "tools" human beings employ to build a view of the world they inhabit in together.
- Language is the supreme human "psychological tool" making forms of learning, problem solving and acquisition of skills possible. Vygotsky recogniised language as primarily a means of communication. The concepts and relationships captured in language are transmitted and acquired in a social medium.
- Children learn by imitation. Interacting with adults and peers in co-operative social settings gave the young learner ample opportuniity to observe, imitate and subsequently develop higher mental functions.

Thus the central concepts of the socio-cultural perspective give currency to language as a tool by which we communicate with our thoughts and mind in learning. Language enables human beings to interact as well as relates experience and knowledge to others (Vygosky 1993).

[82] Vygotsky 1996:12ff.
[83] *Ibid*.
[84] Phillips and Soltis, 1998:59.

is language, the supreme human "psychological tool." This is employed in making higher forms of learning, problem solving, and the acquisition of many skills. Vygotsky believes that these tools make up the culture, and they are the context for interaction in the social structure. Moreover, it is an obvious fact that these tools have tremendous influence on development and socialisation of children.

The critique of Vygotsky is presented in his book *Language and Thought*. In this work, Vygotsky discusses Piaget's theory in a critical manner concerning the "egocentric thought" of the child. Piaget argues that the child is autistic from the beginning, before he becomes social but when this occurs then the egocentric thought looses its function. Vygotsky, on the other hand, sees initially speech as social and comes along side the activities of the child. Consequently, this egocentric speech is translated into the internal sphere, and this happens before the activity. This process offers the child opportunity to utilise speech as the basis for planning, self-awareness, and control.

Essential to the question of development is Vygotsky's concept of a zone of proximal development (ZDP).[85] According to Vygotsky, this is a distance or space between what a child can perform by herself and what a child can accomplish with grown-ups and experts by her side.[86] Adults or older peers' guidance is a necessity. In other words, the zone of proximal development provides that development takes place in transition, that is, between the social (aided development) and the psychological (independent development). In contrast to Piaget, then Vygotsky was not only interested in what a child could be on his own, which is a "static" indicator but also what a child could accomplish with the utilisation of others' guidance. What goes on in the ZDP of the child can be characterised into three processes: (1) There is interaction between parties whose knowledge and expertise are not of equal level. (2) During the interaction process, the child internalises a transformed version of the interaction. (3) The result of the transitions in ZDP is the child's ability to act and rationalise independently. Hedegaard's comment regarding Vygotsky's use of the concept of internalisation is worthwhile:

By internalisation, Vygotsky did not mean copying, but transforming the external interaction to a new form of interaction which guides the child's actions. Internalisation does not directly mirror the external social relations; it is a transformed reflection. On the intersubjective plane, it can be understood as an interaction that the person has made. It still is a kind of interaction, but now the child takes all the positions in the interaction; i.e., the regulating as well as the action role.[87]

Here again Vygotsky and Piaget differ as to how accommodation takes place. Piaget believes that the child constructs his own new knowledge, understanding, and skills. It is a view which states that there is some struggle that goes on in the child's attempt to accommodate new experience to existing knowledge. The construction of new knowledge is continuous with old knowledge. Vygotsky, on the other hand, sees the critical part of the child development as taking place within the mutual interaction with adult. The accommodation occurs in the social interaction, not within the child's mind. Consequently, the development of the child is first "collective" before it leads to the "individual development" of the person. This social setting learning is what made Vygotsky popular at home, and later in the Western world. As Phillips and Soltis point out, Vygostsky's concept of social learning through imitation had been central in the social learning theory propounded by Stanford psychologist Albert Bandura. Bandura innovatively gave a new label of "modelling."

---

[85] Cf. Kutnick (2001:77ff.). This study deals with the interpretation of ZPD in the school context. The paper discusses the historical background of the theory and its application in the real world of relationships and educational contexts.
[86] Cf. Phillips & Soltis, 1998:55ff.
[87] Hedegaard, 2001:16-17.

Situated learning theory

Lave and Wenger's Situated learning theory in their book *Situated learning: Legitimate peripheral participation* (1991) also illuminates and widens the spectre of learning method, as well as the content. Similar to the social theory of Berger and Luckman, the situated learning theory views learning as a change in an individual's relation to his material and social world through acquiring competence with psychological tools/artefacts in situated practice. Some of the central concepts in the theory of situated learning propounded by Lave and Wenger are "social practice" and "production." The method of learning is conceptualised as participation in practice. The concept of "peripheral legitimate participation" is employed to denote a change in status as a result of learning. This change occurs as the learner emerges from being peripheral to being integrated into the social process of becoming a fully accepted competent participant in a community practice.

    Learning viewed as situated activity has as its central defining characteristic a process that we call *legitimate peripheral participation*. By this we mean to draw attention to the point that learners inevitably participate in communities of practitioners, and that mastery of knowledge and skill requires newcomers to move toward full participation in the sociocultural practices of a community. "Legitimate peripheral participation" provides a way to speak about the relations between newcomers and old-timers, and about activities, identities, artefacts, and communities of knowledge and practice. It concerns the process by which newcomers become part of a community of practice.[88]

    Thus, according to Lave and Wenger's approach, the situated practice and institutional traditions become "concrete aspects of learning." Though in a way the relation between institutional practice to both variation in subjects and variation in traditions is still missing, and also how this variation affects learning. Learning theory should take into serious consideration the learning context and the participating persons in the community practice.

Classroom in the sociocultural theorists' view

Researches through experiments have translated the sociocultural perspective into classroom practice. Nuthall provides us with the names of a number of scholars who have participated in the so-called "community of learners" projects.[89] It is an attempt which gained support especially from the knowledge that the social process has a role of influencing the cognitive process. This view, therefore, makes it imperative to creating a social process in the classroom that models effective cognitive processing.

    Adherents view learning as occurring through apprenticeship, and it entails the gradual transformation of the identity of a learner from that of a novice into an expert. In order for this to take place the classroom needs some kind of transformation to occur. Among other things the teacher is advised to model the role of an expert, he should be self-motivated and an intelligent learner. This would encourage the learners to take the role of "learning apprentices." Furthermore, in order to build a community of learners, two things need to be taken into consideration: (1) Officers of education should define, model, and practice the activities of learning; (2) They should also institute the norms of practice which will structure meaningful participation in communities of learning. In the experimental studies that were

---

[88] Lave & Wenger, 1991:29.
[89] Brown (1992,1994); Brown, S, Rutherford, Nakagawa, Gordon & Campione (1993); Brown & Campione (1994).

done by these researchers the model of "teacher-learning groups" was practised and the norms of co-operative collaboration was utilised. The consequence of this kind of the classroom modelling was students maturing to a certain level as members of a community of "research practice." What followed was a successful imbuement of the manner of knowing the cultural practices the discourse patterns and belief systems of scholars.[90]

Roger Säljö and the concept "practice"

Let us consider the work of a Swedish researcher called Roger Säljö as he has also contributed to the sociocultural perspective in recent years through his publication of a book entitled *Learning in Practice* (2000). This work provides us with a detailed analysis of some of the most important central concepts in the sociocultural perspective. Säljö, in a unique manner, deals with learning as a social phenomenon. His presentation of the concepts in his book would make us become familiar with some of the other aspects that are taken for granted by other theoretical perspectives. Here, the "practice" concept and the three aspects associated with it are pinpointed. By "practices," Säljö means human systems for acquiring certain goals, serving certain functions, and producing certain results. These systems utilise capabilities that are distributed across participants and are also spread over tools. Individuals learn by participating in the practices. But not only do individuals learn in the practices, practices also develop as human beings learn. This takes place by developing new or improved tools. Language is one of the most important features of these tools. Through the language tool, reality is mediated in the practices that are above all "discursive practices."

Practice in school

In Säljö's understanding, the school and the institution of learning is considered as a particular place where social "practice" of learning takes place. In other words, the school offers a particular kind of social practice and possesses its aims, as well as its distinctive characteristics of carrying out measures, making judgements, and "ways of including and excluding behaviour."[91] Students are said to participate in the social practice of the school. Säljö in his theory does not adhere to the theorists who are interested in studying differences in students' heredity or differences in the environment. He, like the sociocultural theorists, believes in differences, which is caused from variations in the practices or differences between the conditions for the individuals. According to Säljö, the school and institutional learning practice differ from other forms of practices in that learning is not "a by-product, but the very aim of the practice of institutional learning."[92] Students' learn in school, which equip them with certain capabilities that enable them to use these capabilities in other practice. Known as decontextualisation, this idea is a well-known problem characteristic with institutional forms of learning, according to the sociocultural perspective.[93] *Differences in learning could be explained as resulting in variation between the practices.*[94]

It could be seen from his analysis that Säljö leans heavily on studies of historical and cultural variation in the practices; the focus on his learning theory is also embedded in it. One could argue that his work was built on the historical comparisons in the Vygotskian

---

[90] The book edited by Mariane Hedegaard *A Cultural-Historical Approach. Learning in Classroom (2001)* Using Vygotsky's ideas, the authors contribute to the cultural historical study of school children's learning.
[91] Cf. Merton 2000.
[92] *Ibid.*
[93] Cf. Säljö, 2000:35ff.
[94] *Ibid.*

tradition, as well as the more recent cross-cultural anthropological tradition. Obviously, the empirical examples he provides in his work are about variation between practices and variation in the defining features of practices. The fundamental notion that appears in his work concludes that learning or performance varies when the conditions vary. The connection of his work to these perspectives enables us to visualise the outstanding central concepts at work. We will focus on these concepts briefly and show their importance to variation that results in learning.

Three Aspects of Practice

Three aspects of practice that Säljö discusses in his work are: 1) the use of tools of the biological man and his development from biological being to a social being; 2) the social nature of learning, which provides capabilities to people of diverse society; and 3) the primacy of mediated, discursive character of practices.

"Tools" as a concept in the sociocultural perspective also appear in Säljö's analyses. According to Säljo, although human beings evolved through a biological process, his subsequent advancement or progression has been sociocultural. The biological ancestors who lived so many years ago before us are not different from us except that the progression in development until now has made us emerge as powerful social beings due to the invention of tools. Tools can be "real," that is material which ranges from a hammer or wheel to spaceships, etc. The tool concept can also be denoted conceptual such as science, or linguistic, such as human languages. Thus Säljö, like other theorists such as Bowker and Star,[95] consider language as an important communication tool which aids learning. It is a significant tool in arranging and organising human activities in a more concrete sense. Practices of categorisation entail part of the concrete ways of assigning and handling the world, including social problems.[96] Through language we experience and share knowledge in the form of experiences to others. It is a unique way leading to creation and the communication of knowledge.[97]

Distributed capability or distributed nature of learning and capabilities indicates that practices are participating by people together with other people with the use of tools. In the social world, therefore, the individual is not the best choice of description when one wants to characterise what people do and what they are capable of doing. The team-in-the-setting-making-use-of tools are, according to Säljö, frequently a more relevant unit of description than single individuals.

The third aspect of the social practice is the "discursive practices." It is recognised as constituting a vital part of research orientation, and in Säljö's presentation emphasis is put upon this. In order to become expert in a practice, a person has to master its discourse. The same applies to the use of conceptual tools which one needs to take part in its discursive practice if one wishes to master them, especially where there is a need for these tools. These discursive practices, that in which an individual takes part in, shapes and moulds this individual's comprehension of the world around him. They not only delimit the phenomenon by pointing to its features, but they also make certain aspects figural, allowing others to stand back. Reality cannot be experienced as such; one encounters and grasps a

---

[95] Bowker and Star 1999.
[96] Douglas 1986; Mehan, Mertweck, & Lee Meihls 1986.
[97] Språket är en mekanism för att skapa mening som det mänskliga språket utgör. Språket är en mekanism för att bildligt uttryckt lagra kunskaper, insikter och förståelse hos individer och kollektiv. Genom att tolka en händelse i begreppsliga termer, kan vi jämföra och lära av erfarenheter (Säljö, 2000:34).

discursively mediated reality. The practices one takes part in and discourse employed can never be transcended. "Truth" as a result becomes relative to the criteria employed in distinctive discursive practices. Truth then vanishes as a transcendental category. Thus in the analyses of Säljö we find that the central concepts of the sociocultural perspective are applied to the school context, a brilliant attempt which, in my opinion, differentiates his work from all others. As the concept "language" appears in his descriptions, giving it a central role, so have some researchers also furthered the language concept, making it another separate perspective altogether.

The Language Focused Approach

This notion is derived from the writings of those who had earlier on pioneered the sociocultural perspective of learning. In fact, one could honestly say that the language-focused concept is a subset to the sociocultural perspective, since writers constantly refer to the names of Vygotsky, Lave, Rogoff, and many others. We need not discuss it here, but as we want to make an overall review of the learning perspectives we ought to deal with it, probably in brief.

According to Nuthall, three distinct kinds of language-focused perspectives of classroom process can be identified. They are:

(1) Those that utilise methods and concepts of analysis enshrined in linguistics and in sociolinguistics.
(2) This variant is rooted in the sociocultural perspective and bases its views on language as a cultural artefact.
(3) Those who discern a specific linguistic genre as possessing central importance in cognition.

Sociolinguistics: "talking" knowledge

The first of these approaches appears in the work of a research "Discourse group" based in the University of Santa Barbara. This Classroom Discourse Groups'[98] interest is focused on comprehending how daily experiences in the classrooms are constructed by members in the course of interactions (i.e., verbal, tasks). They are also attentive to how these constructions affect what learners acquire, accomplish, or learn in schools. Other activities such as role playing, taking responsibilities, norms and mutual expectations between learners and teachers, are observed. As expected, in all these, language is the medium employed to construct and provide understanding of the purposes, events, classroom tasks, and the content of curriculum of all the activities that take place in the classroom.

According to the Santa Barbra Research Group, "two intersecting dimensions" are seen as the result of what transpires in the classroom of learning: (1) The social life of the teachers and learners is developed and played out through negotiations of roles and status. (2) There is a gradual development of knowledge of the curriculum with its concomitant awareness of opportunities and processes of learning. Fresh ideas of knowledge or what is accepted and understood as knowledge comes to the fore as teachers and learners come into contact with each other. These kinds of knowledge (be it academic or social) are executed into being by a range of discursive practices that appear in teachers' and students' discussions at

---

[98] Cf. the work of Green & Dixon (1993:231ff.).

school.[99] This approach of learning, therefore, considers teachers as possessing enormous responsibilities in that they communicate or "talk" knowledge into the learners. Thus, the research group's greatest contribution, as Nuthall notes, is that they point out the critical function mutual understanding has in communication.

Lemke[100] gives a new look to the sociolinguistic approach when he showed in his work that classroom language could be received "semantically" as well as "sociocultural" analyses. His main concern was to investigate how underlying meanings come to together through human interaction. For him social interactions make it possible for individuals to create meanings. Daily conversations enable individuals to make connections between one another's experiences and this is a practical way to make meaning.

Language as cultural artefact

The second line of the language-focused perspective is based on the work of Vygotsky, Bakhtin and Leont'eV [101] and it deliberates on the learner's ability to master the utilisation of cultural tools. It is in and around the social context that learners develop their mental capabilities, such as thinking and memory through the use of cultural tools. Such tools as language, science, and art works contribute to the development of the mental faculties and skills. The learner inherits the knowledge and concepts enshrined in the tools through the utilisation and mastery of cultural tools. Language genres are the most vital cultural tools within the educational establishment. These set of genres or discourses contain the concepts and thinking of the different areas that make up the curriculum. As the learner studies the curriculum that pertains to his subject area, he not only becomes aware of the concepts and principles, but acquires a special way of perceiving, thinking, and classifying the experiences which are inherent in the language. Imbibing the symbol systems makes a further acquisition; the technology utilised by the experts in his subject's area. It is to this extent that learners who engage in the discourse, be they technical or cultural practices of any particular area in the curriculum such as agricultural science, history, or mathematics, is regarded as "semiotic apprentices".

Linguistic genre as significance in cognition

Classroom researchers who employ this specific approach contend that the narrative genre is the fundamental organising structure that determines the manner learner's process, comprehension and recall to memory their experiences. The outstanding scholars of this thinking are Hicks and Egan,[102] whose recent studies have unveiled this perspective.

Hicks, for one, is of the opinion that the structures of the learner to be thought are a copy of the discourse structures of the classroom language. He argues that the learner in his day to experiences imbibes the discourse forms that are provided to him by the social world through social interaction. This, therefore, serves as a means for his sense making in the world (1993:131).

Hicks disagrees with other researchers who contend that several different technical and specialised genres mould the thinking of students. In contrast to this, he suggests that only one genre, that is, the narrative discourse moulds the learner's thinking.

---

[99] Cf. the article of Nuthall (1997) for empirical examples.
[100] Lemke (1990).
[101] Vygotsky, Bakhtin and Leont'eV (1981).
[102] Hicks (1993) and Egan (1989, 1993).

The structures of the narrative discourse, in it are the primary intellectual tools which equip learners to make sense of their own world. Learners who have access to the narratives use the symbolic means it offers for structuring events into a meaningful and comprehensible whole. The context of the story is important as it presents different events form and significance.

Children become aware of narrative prior to their commencement at school. As soon as they begin to interact with grown-ups that they have inherited oral culture from they begin to utilise the narrative structures they have acquired in the community to structure and comprehend their world. The school performs the job of expansion, and the formalisation of the forms of discourse which the child has already acquired. The implication one ascertains from Hicks, according Nuthall, is that the discourse structures which the child has acquired gives the means by which the child build up new cognitive structures which consequently leads to his intellectual development. This perspective which has been discussed therefore views classroom discussion *as the creation of a narrative*. It is from this line of thinking that many writers, including Schank,[103] have stressed the importance of language acquisition through social interaction.

Phenomenography and Variations Theory

In his attempt to contribute to our understanding of how learners acquire knowledge, Ference Marton in Gothenburg University developed the "phenomenography" research method. The approach has since been employed to conduct research all over the world. Recently, another approach has been developed in the area of classroom study. This latter perspective connects the model "phenomenography" and the manner variation contributes to the study of the content of mathematics and it has been named "variations theory." The author herself states that the variation concept that is employed in her work is intrinsically connected to the theoretical perspective of learning that is newly presented in the work of Merton, Bowden and Booth.[104] The theory is new and the concept or the building blocks ought to be developed through empirical testing to be able to have it related to other well-known theories.[105]

Marton's contribution to this research approach in the classroom was first published in an article entitled "Phenomenography– Describing Conceptions of the World Around Us." Here Marton argues and puts forward in his paper in favour of research, which purposes in "finding and systematising of forms of thought in terms of which people interpret significant aspects of reality."[106] The approach is said to identify, formulate, and deal with certain kinds of questions of research that relate to learning in the context of education. When researchers employ this method of learning to embark on research they are mainly interested in how to identify and describe the different manner in which individuals experience and comprehend certain phenomena in the world. *Phenomenography is therefore a method of description, analysis and comprehending of experiences.* Earlier in his first paper Marton distinguishes between two perspectives which he considers as fundamental to this research approach. According to Marton,"from the *first-order* perspective we aim at describing various aspects of the world and from the *second-order* perspective (for which a case is made in this paper)) we aim at describing people's experiences of various aspects of the world."[107] Though in his pioneering work he argues in support for the second-order perspective, later he advocates for these two perspectives since in his understanding both perspectives are

---

[103] Schank (1990).
[104] Cf Bowden & Marton, 1998; Marton & Booth, 1997.
[105] Runesson, 1999:312.
[106] Marton, 1981:177.
[107] Marton, 1981:177.

complementary. This approach, therefore, is not only concerned with people's experiences, but also the "variations" in people's way of comprehending or perceiving the world around us.[108] A later clarification of the concepts in the approach has been made through successive publications which unveil the method to be entailing a model for learning and awareness.[109]

Toward experiential description of learning

The method, simply stated, is interested in people's experiences of the world around us. And that is what researchers who utilise this approach in classroom learning also comprehend. "Experience" could therefore be considered as a very important concept central to phenomenographic studies. Borrowing from Gurwitch's[110] concept "field of consciousness," Marton and Booth[111] illuminate on the concept "experience" in relation to learning. They contend that the way one experiences something or a phenomenon is quite related to the manner one's awareness is structured at a certain moment. In order to experience an object or a thing, for example, "matter" or "force," certain properties of the object have to be simultaneously be discerned and kept in the focal awareness. The sequence of experience is: 1) an object must be discerned from its setting; 2) the properties of the object must also be discerned as well; and 3) then the properties of this object must be related to each other and to the whole. Marton and Booth argue that a person's way of experiencing an object is the result of the aspects of the object that are *discerned* and the relation between them that are kept *simultaneously* in the person's focal awareness. Experience or learning for that matter is:

Formulated in terms of the structures and dynamics of human awareness. A way of experiencing a phenomenon, a situation a problem etc., is thus defined as the aspects of the phenomenon, the situation, the problem etc. that are discerned and are simultaneously present in the individual's focal awareness. Those aspects that are discerned represent different dimensions of variation in awareness.[112]

The learner and his learning

In his article published in *Forskning om Utbildning* (1992), Marton indicates that the ontology phenomenography rests on is non-dualistic, and that knowledge is neither objective nor subjective but both, suggesting that the object and the subject be related internally. What a child does as it emerges into this world is to become a part of the world at the same time that the world tries to incorporate him. Knowledge is then gained through the relationship of interaction that occurs between the learner and his world. Consequently, knowledge could be considered as both personal and collective, which means it is partly experienced by learner and partly beyond the learner.[113]

Marton's definition of learning is a matter of perceiving, conceptualising, experiencing, or comprehending something in a different way from previously.[114] It is a change, which comprises a learner's ability of experiencing something in the world. Learning

---

[108] Bowden (1996: 60 ff.) points out differences of focus that usually exist nevertheless they are also important variations.
[109] Marton & Booth, 1997; Bowden & Marton, 1998.
[110] Borrowing from Gurwitch (1964).
[111] Marton and Booth (1997).
[112] Runesson, 1999:314.
[113] Cf. Marton, 1994.
[114] Marton & Both, 1995.

also implies that more or other properties of the object are perceived and related to the whole, or that properties are perceived differently. Simply put, *learning is a change in the structure of the individual's awareness.*[115]

What Marton would like us to understand about children is also very important. According to Marton, children's experience is not different from other forms of knowledge; it consists of a whole, an understanding, which can include skills, facts, and intimate knowledge. Children experience something on a social, intellectual, and emotional plane at the same time. Things are perceived in a constant way. Knowledge possesses a social character; because of this the child's experience also takes place only in a social context. Knowledge can, therefore, be understood as context dependent. Experience constrains people to seek for understanding, and as experience changes a new understanding is sought by people. This explains why people are constantly involved in interaction with the world around us. Learning is inseparable from development. Learning leads to development, and as we develop we learn.[116]

Variation in learning

The variation perspective is intrinsically connected with the way learners perceive an object or a thing. In general, individuals succeed in keeping all-important properties of an object in their focal awareness simultaneously. This suggests that there are variations in the manner in which a phenomenon is experienced. Already Marton and his followers, through research, inform us that within the tradition there are numerous accounts of the different manners in which learners experience various phenomena in the world. Marton & Booth[117] interpret the differences in experiences as resulting from differences in individuals' capabilities to visualise or perceive critical properties of the phenomena in question. Pramling discusses that the whole idea of phenomenographic research is variation of people's way of thinking:

The aim of phenomenographic research studies has been, from the start, to describe the variation of ways of thinking about specific phenomena. The result, according to Marton,[118] the "outcome space," is a set of different ways in which a group of individuals understand a specific content. Since all people have different experiences, phenomena in the surrounding world will appear different to different people. The ideal outcome space, all possible ways of experiencing something, constitutes the phenomenon as such. The different ways of thinking are considered as qualitatively different, partly dependent on the difference in the conceptualised acts and partly depending on types of content. Thus in some cases one way of thinking is judged as more advanced than another, while in other cases the conceptions are regarded as horizontal and equal but still qualitatively different.[119]

Three Aspects of learning: Variation, Discernment and Simultaneity

Phenomenography way of doing research has it that before an individual learns something, these three aspects have to function in the individuals' mind to make learning possible. We have already asserted that from phenomenography point of view, the object in the first place must be discerned from its context and certain aspects of the objects must be discerned and

---

[115] Runesson, 1999:315.
[116] Cf. Pramling, 1996:84.
[117] Marton & Booth (1997).
[118] Marton (1981).
[119] Pramling 1996, paraphrased from Uljens 1989.

related to each other and to the whole.[120] According to this method, to experience an object and its colour, for instance, entails an experience of variation in some respects. It follows that to discern those aspects of the object one must relate it to potential dimensions of variation. The colour of the object, generally understood, suggests a value in colours dimension. To discern the object in question's colour, a previous experience of other colours (red, yellow, white or blue) must have been experienced.[121] Nor is this all; the substance of the object, say, a clay form, a transparent object, etc., must also have been experienced. Consequently, the manner the object is experienced as well as the meaning accorded to it is a function of the dimensions of variation through which it can be observed. It goes without saying that to acknowledge what something is, one should be aware of what that something is not. It could be maintained that, discernment; simultaneity, and variation are all related to each other in a logical manner. This is what Marton and Booth,[122] as well as Bowden and Marton,[123] would like us to comprehend or argue, that awareness possesses a structure and the manner any object is encountered or experienced can be analysed in terms of the structure and organisation of the awareness at a given time. Structures enshrined in different understandings of different phenomena are intrinsically connected to the meanings one accords to these phenomena.

These three aspects of learning: "variation," "discernment," and "simultaneity," have played a major role in Ulla Runesson's dissertation entitled *The Pedagogy of Variation* (1999). Runesson used these central concepts, especially the notion of "variation," to deepen her studies of the different ways teachers handle the content when they teach fractions and percentages to pupils. Runesson's work was the first to support and recognise the notion of variation as a theory. She applied this theory to teaching in the classroom. In her work she paid attention to the different aspects of the content of teaching, how these different aspects of content are "focused" upon or "thematised," what aspects are left unfocused, and whether these focused aspects unveil what she calls "dimensions of variations," or not. In a lengthy summary of her dissertation one gets a glimpse of her results:

Let us imagine two teachers, both wanting their pupils to develop an understanding of the concept "square." Let us also imagine the first teacher dealing with this by using a picture of a square, first focusing on the size of the angles, pointing out to the pupils that all four are "right angles". Then she focuses on the four sides and the relation between them (all four sides have the same length, and the sides are parallel). Finally she gives them the mathematical definition of a square. Now, in contrast to this, let us imagine the second teacher focusing on the same aspects (the angles, the relation between the four sides), but doing it by introducing a variation of these aspects. When focusing on the angles, the teacher compares the picture of the square with one of a rhombus. She points to the angles in the rhombus and contrasts them with angles in the square. By doing so, she exposes a variation of the size of the angle. She opens up for a variation in that dimension. A right angle can then be experienced by the pupils as one value in the dimension "angle-size." Correspondingly, she can vary other aspects of the square and thus open up for other dimensions of variation: the length of the sides, the number of sides, different squares (i.e. the length of the sides of the squares are different), etc. She can also focus on other aspects of a square, such as the occurrence of squares in everyday life, pointing out that the flooring material is made of shapes, the chessboard has 64 squares, the six sides of a dice are squares, etc. By doing this, she opens up a dimension of variation of different squares.[124]

---

[120] Cf. Svensson, 1984.
[121] Runesson, 1999.
[122] Marton and Booth (1997).
[123] Bowden and Marton (1998).
[124] Runesson, 1999:317.

According to Runesson, teachers handle content by allowing some aspects of the square to become invariant, while others become varying. This procedure enables teachers to accomplish successfully "a space of variation," and this refers to the meaning of a square that is given to the pupils to digest. Borrowing Patrick's[125] terms, the teacher "moulds" an object of study that is given to the pupils to experience. The study object serves as a potentially experienced object and does not become the real object, that is experienced by the learners. Here in this example, Runesson illuminates clearly her work, which utilises the concept of variation as a function in the handling of mathematics content in the classroom. The examples of the two class teachers mould two diverse study objects. As the second teacher presented her teaching, she handles the square concept such that learners get the possibility to potentially experience the meaning of which aspects can vary and that which should remain invariant. As this is done, it enables the characteristics of a square to be visualised as against that which is not a square. Runesson's study, like all other phenomenography studies, concentrates on identification and description of meaning given by its informants. But her work goes a bit further than others do, as she extends her analysis to explore the *expressed* and *potentially experienced* meaning rather than only the expressed meaning. She employs analytical tools such as the focused or the non-focused aspects, simultaneity, and variation to unveil the communicated meaning accorded to the learners in the process of teaching.

Currently Runesson has gained recognition in her contribution to "Variation theory," which her name is associated with. She has also made enormous contribution to how teachers handle the content of mathematics with her use of the phenomenography method for analysing the information gathered from her fieldwork. It seems to me that her perspective could be a bit attractive to scholars if she could utilise some psychological concepts to deepen her developed theory of variation. I would suggest that Runesson's perspective could employ some mathematical concepts to represent some of the propositions she makes at the end of her studies. She could, for example, find an appropriate mathematical formula to represent variation theory as a whole.

But my purpose is not only to examine these old perspectives, but also to go further and propose one additional theory, which I believe, will be of use to the learning theories meticulously examined above. What is the theory of blood language?

The Theory of Blood Language

These perspectives analysed above recognise the importance of language in the learner's acquisition of knowledge both in the classroom or elsewhere in the society. This recognition is great, as it enables educators to become serious and attentive in the child's use of language during the teaching process or after. It is acknowledged that many writers have developed all these perspectives not only because the teacher or instructor uses language to communicate knowledge to the child, but also because without proper language acquisition the child may not comprehend the lessons being taught in the classroom. *What the theory of blood language proposes is that it is only the child's use of the language that is fixed in his/her blood that will guarantee a successful learning experience and overall performance in the society. The maximum fulfilment of his goal depends on the use of this language that is naturally part of him.* Moreover, successful learning that leads to originality in thinking, sharpness in carrying out responsibilities, and accuracy that is required by such sciences as Mathematics, Physics, Chemistry, and etc., can only be obtained when the learner uses the language that flows in his blood.

---

[125] Borrowing Patrick's (1997).

The other theories do well when they stress on the cognitive aspects, the social and cultural perspectives, and the notion of practice or rehearsal in the classroom and outside the classroom. They have mentioned the use of artefacts, and language itself being explained from different vantagepoints. Because they have taken it for granted, they do not go further to stress this important need of the learner making use of the language that he is born with. Whenever we talk about the language of the blood or the "first language," as usually denoted by many, we think about making learning easier and so think first about: (1) Comprehension, (2) Description and analysis, (3) Originality in thinking and creating, (4) Accuracy and sharpness, and (5) Carrying out responsibilities in the society.

By this design I shall offer points in the form of axioms or propositions and then use observation, empirical materials, and reason to prove them. Let me premise this axioms with definitions, for as Aristotle saw this clearly, "the basic premises of demonstrations are definitions."[126] With the definitions and the premises, we can then move further to deduce the theorems for our new theory.

Definitions

Definition 1

By language I understand the method of human communication, either spoken or written, consisting of the use of words in an agreed manner. It consists of a style or the faculty of expression and the use of words or signs.

Definition 2

By the language of the blood I comprehend that it is the language that an individual is born with which, if it were to be dissected, it could be described as flowing in his veins together with his blood. This language is "fixed" in the individual, and it is transmitted to successive generations through procreation.

For it is understood that both parents of the child, or at least one of the parents speaks this language. It is the language that naturally comes out from the mouth of this child as he/she begins to speak. Because the parents, by bringing up the offspring, do not consider any other language as theirs except that which they have inherited from their own parents, and comes naturally from their lineage. It is also understood that both parents have consciously and through effort tried to propagate this language to the child and believe themselves that it could be the language that the offspring should be educated with. And so the offspring, by acknowledging this language, neither have to suffer or undergo some difficulties in comprehending this language.

Definition 3

The learner is the individual who through agreement either between his parents and the education authorities or between him and the authorities directly undergoes an apprenticeship. The individual is taught by well-trained teachers who help with communication to both impart knowledge to the individual and he, the teacher, reciprocally gained from this interaction.

---

[126] Posterior Analytics, in *Works*, ed. by W. D. Ross, Vol. I, 1928 quoted in Cohen & Nigel, 1949:232.

Educators usually consider education to be a kind of apprenticeship where the individual learns to practice, and through this gain knowledge. This knowledge that the learner acquires are benefits to the state and to the learner in that while the former spends money on the latter, the latter intend to utilise this knowledge to serve the state and in return gain something for his livelihood. Learning itself is knowledge acquired by study, either independently or through the help of an institution.

Definition 4

Comprehension of something is the meaning the learner makes of what is being communicated to him by the instructor or teacher. It is the perception of the meaning of what is being asserted or what one is observing.

Definition 5

Description is the manner of stating the characteristics of something and through this be able to portray how that thing being described looks like or appears to be in reality.

For by this word one is able to carry out meanings from something one has perceived to another person or to oneself. It can be presented in a written form or orally. As description involves analysis of some kind, it is very vital for the learner to be able to do that. Examinations are usually performed through this means, and many tasks that the graduated learner undertakes are carried out through this means, either by description, or analysis, or both.

Definition 6

By originality I mean the power of thinking or creating creatively. It denotes newness or freshness, which concerns the idea of bringing something new into existence or discovering it.

Definition 7

I understand accuracy to be exactness or precision resulting from careful effort. It denotes the degree of refinement in measurement or specification.

By this word we understand the manner one carries out responsibilities without making unnecessary errors. By this we also understand the human effort of persevering and trying all means to be precise and exact. It does not come automatically, but with effort.

Definition 8

Sharpness or alert means being attentive/awake and able to catch the meaning of any sign or movement made by another party. It means the individual is not dull, but instead bright.

Axiom and Propositions

Axiom 1

Whenever learning is contextualised into the locality of any learner, the major burden involved in how to imbibe knowledge and systematise them are tremendously reduced, which consequently make easy comprehension, analysis and description, originality and alertness to occur to the learner. This accords the learner with a higher degree of capability to take responsibility later in the society.

Axiom 2

The utilisation of the learner's own blood language in educating himself/herself is one such adaptation of learning to the context of the learner.

Axiom 3

Accordingly, the use of this language that is fixed in the learner is the sole language that the acquisition of learning and the carrying out of assignments in institutional context should employ.

Comprehension—Proposition 1
Theorem 1

If the learner uses the language that flows in his veins/ blood, I say that it will quicken his comprehension ability (perception) in a high degree, and the time spent on revising for examination or rehearsal and other purposes in the context of learning will be greatly minimised.

Most of the learning that takes place in the classroom and in the society occurs to the learner through understanding something that is being taught or going on (Def. 4). This child has to absorb the lessons being taught in order to reconstruct it and be applicable or used by him (Def. 3). Here we find out that unless the individual has been born and bred in the language (that could be as well counted as fixed in him Def. 2), though he may follow the lessons or the programme, he may still be handicapped in his comprehension, this being the vital part of learning. But if the language that is fixed in the blood were what is being utilised to conduct this programme or learning, he would not have to strife in order to comprehend what is taking place. On the other hand, supposing the language is a borrowed one, usually such learning has lapses, as the learner first has to comprehend the language being used as a medium and at the same time follow and comprehend the content of the lessons. This becomes a hectic problem. No wonder people that have that language as their second language usually have problems regarding their progress during and after learning (Prop. 5). If the language that is being used to conduct the teaching were the learner's first language (Def. 2), he would have a higher degree of comprehension that would also enable him to follow the lessons without any difficulty. This latter learner would also spend fewer hours in preparing for examinations or revising the lessons.

Description—Proposition 2
Theorem 2

Description is a required talent of the learner's learning capability, and the acquisition of this talent is correlated with the use of an individual's employment of his own blood language which if denied of its use, can make him handicapped.

To describe something one needs to employ language (Def. 1 & 2) and its demands entails the use of different phrases, complex grammatical laws, which consist of present tenses and past tenses, and etc. An individual who is to describe something, that is, if he employs his own language, does this without any difficulty. It flows naturally like water flowing from the top of a mountain. A learner that has studied the language which is not of his own may describe, but with difficulty. Though he may be able to employ all the proper tenses, he will still lack the courage and beauty to describe which will be able to help someone follow what he is describing or be put down as a model. No wonder one becomes afraid to use the description of someone who does not have the language as his own for a police report. Learning is gained through comprehension (Def. 4), and being capable of describing what has been understood is another important facet. If one fumbles, the whole learning loses its meaning. It is therefore necessary to be able to describe well with one's own language what one has stored in the memory. People that use the language in question as their second language seldom achieve perfect analysis and description. This affects their productivity in terms of writing books, projects, and creating something.

Even to gain proper acquisition of knowledge the sensory experience sets the problem for knowledge, and just because such experience is immediate and final it must become informed by effective analysis before knowledge can be said to take place.

Originality—Proposition 3
Theorem 3

Originality in thinking and performing tasks is directly related to a high degree of use of the language that flows in the blood of the learner, and if denied of its use, could affect his overall intelligence capability or performance level.

In learning, one is expected to comprehend and make use of the knowledge one has imbibed (Def. 4). It may be that this knowledge that one has acquired will be able to aid one to fulfil a responsibility or work with an area that needs development. To use another person's language makes learning understood as something that is associated with things about that person. It does not make learning becomes part of the learner. This occurs with the learner, as sometimes if it is language speaking, when he wants to speak another's blood language this individual has to change his voice and try to pretend like one of the people who is speaking their language. It becomes an imitation that is not well adapted to this individual's cultural context. In other words, he cannot make this imitation part of him that can be successfully developed to originality. To use one's own blood language in acquiring knowledge, it makes the knowledge becomes part of the individual's self (Def. 2). From this comes originality that is already reinforced by the fact that the language flows in his blood and the learning becomes the natural acquisition of all that are being imbibed (Def. 6). It is this kind of originality that aided many Europeans to think independently and be able to invent, discover, write books and poetry, and many things when they all shifted from **Latin** to their various blood languages. It pushed away ignorance and naivety. It is the blood language that enables the learner to be original in his thinking and perform tasks without much difficulty. It makes the learner shift his/her interest on studying the things around his vicinity and become obsessed with developing his locality more than ever.

Accuracy and Sharpness—Proposition 4
Theorem 4

If the learner is not given the opportunity to utilise his own blood language, it will in a high degree affect his performing tasks with accuracy and sharpness, and this will be seen in areas that have to do with the Sciences and Mathematics.

In the science subjects (i.e., Mathematics, Chemistry, Physics, etc.), accuracy and sharpness are needed to be able to carry out certain tasks (Def. 7). The learner that uses another language that is not part of his faculty is always beset with problems concerning accuracy. To strife for perfection becomes a matter of difficulty. No wonder in the days when Latin was being used, one could count the number of scientists in the then world before the Enlightenment and during the Enlightenment period. In Newton's time many people could not utilised the brain in solving problems, that is in the domain of science (i.e., mechanics). The accuracy we see with some species at present is only due to their employing *their own blood language* (Def. 2) to the scrutiny of the sciences. It quickens the sharpness of the learner and he is able to perform tasks with much accuracy (Def. 8).

Carrying out responsibilities—Proposition 5
Theorem 5

To be well-enlightened means one uses the language that flows in the blood in educating oneself, and this influence the learner's efficiency level in carrying out responsibilities and assignments, as well as this quickens the overall development of the learner.

With the use of the blood language, all the above advantages will be obtained, and the result is a better and efficient manner of carrying out responsibilities in the classroom and also in the society at large (Def. 2). It is the used of the blood language that has helped many European countries and also those in the Far East to quicken their development. Even any country that intends to use the democratic principles in its society should be able to depend on the blood language. If you import a language, the probability is that people will not have a high degree of interest in absorbing laws written in a foreign language. With the use of the blood language, a learner who has gained a secondary education can easily perform almost all the tasks a university graduate is supposed to do (Def. 4). This can happen, though not in all of the disciplines, at least in the majority of them. Those who have basic elementary education can be reckoned as well enlightened people, especially if these people pursue private independent studies while advancing in years.

Major and Minor Theorems

Let us now prove about the theory of blood language five theorems, which I believe is of vital interest to the learner and those who educate others. These theorems amount to asserting that the learner's state is in shambles if he employs the language that is foreign to him to educate himself. These theorems at once suggest the problem faced by many Third World countries that still utilise foreign languages to educate their children from the inception of their education.

Theorem 6. *All learners who use the languages that flow in their blood in educating themselves are bound to remain original and perform tasks accurately in a sharper manner than those who employ other languages that do not inhere in their blood.*

This follows at once from propositions 3, 4, and 5. It has been a characteristic of many Europeans to see themselves as clever and all other races as not clever. But observation and empirical evidence support our theorem that being able to educate oneself with one's own language and not another's can bring the individual to attain some originality. In this context, observation has shown that in West Africa those mechanics that employ their language during apprenticeship to repair cars are capable of graduating within a one-year duration. They attain originality in building and innovations in some things. At the Kumasi Magazin, in Ghana, these young boys perform wonders such that they could be comparable to many mechanics in Japan or elsewhere in the world. Those who use the English language as a medium of the learning they had received could spend four years at the University and they could either do the same thing or not, depending on how serious the graduate took his studies. If Africans could use their various languages in educating themselves, they could compare themselves to any other species on earth. At the moments they are buried in ignorance, just as it happened in Europe so many years ago when Latin was being used for their education.

Theorem 7. *To be able to comprehend and describe one's learning is the task of the learner, and I say that if the language being employed as a medium of the learning is different from the language fixed in the blood, the progress of the learner will be hampered sooner or later.*

This theorem is deduced from propositions 1, 2 and Def. 3. This theorem explains what is going on in many Third World Countries and Africa, for example. When a child is born and bred in a culture that his own blood language becomes latent, and at the age of five he begins schooling with a strange language, during the first day at school a problem arises. First, he cannot speak his own language well. Nor can he write it, but suddenly he is introduced into another, which he may not be able to master throughout his life. When this is done, his manner of comprehension, power of analysis, originality in thinking, and the manner of carrying out responsibilities all are hampered.[127] The theorem explains why many Europeans,

---

[127] **THE MILLENNIUM HYPOTHESIS**

If each African country could choose a nucleus language that is spoken and understood by the majority of its citizens, and educate themselves with this particular language, it will quicken the pace of their development as well as their capacity to carry out responsibilities.

Corollary:
The hypothesis states further that their swiftness, sharpness, efficiency and their ability to be reckoned with as serious nations in this technological age depends on their being capable to rehearse and imitate properly and to utilise their own blood languages to label even their finished products.

by changing from Latin to their various languages could become well emancipated (Prop. 5). The number of writers increased, poets, scientists, and all others. And moreover, their manner of analysing something became perfect, and sadly enough this has been attributed to their being a superior species. This is not correct. We may as well point out that the problem faced by people who have to forget their own blood language forever is enormous. But as for this, no one will be capable of accessing the Black population in America for what they have gone through, or what might have been misinterpreted, as they do not measure up just because at early stages they were made to use a language that was totally foreign to them (i.e., not their blood language).

Theorem 8. *Under the same supposition as in propositions 1 and 2, if the learner employs the blood language in educating himself or herself, I say that learning generally will become easier, enjoyable and desirable.*

This is evident by props. 1, 2 and 5.

Corollary. Hence, the use of the original language of the learner will make pursuit in education meaningful.

Theorem 9. *With the same conditions as in proposition 4 and Theorem 6, if the learner uses the language that is fixed in his blood, I say that he will make greater advances in the area of mathematics and the natural sciences.*

This theorem is evident by the very propositions mentioned above, that is, props 4 and 1.

Corollary. As these subjects demand exactness and more accuracy, the use of this original language will relieve the learner from certain unnecessary burdens and this will pave the way for the learner to progress with little effort.

Theorem 10. *Let the use of the blood language in educating the learner be the concerned of the learner and the society in question, it will lead to greater advances in development of the vicinity.*

From what have been asserted it will be understood that as soon the society of the learner makes a quick turn of concentrating on the use of the language of the local people, this will call for the development of certain important things in addition. The result will be the initiation of major developments that have affiliation with this particular language. The citizens will be aware of the need of raising standards in further areas of the locality that need development.

       Corollary 1. Hence the employment of the language that inheres in the learner's blood in educating oneself will lead to major progress in the society (by Prop. 5).

       Corollary 2. If the utilisation of the blood language leads to advances in development then this will cause enlightenment in the society.

       Corollary 3. The society that makes use of its own blood language in the education of its citizens will find it successful in raising standards.

Definition 9

By Secondary Law of Intelligence I comprehend the employment of the scientific method to acquire knowledge. It is the approach that uses the systematic manner to gain knowledge in the sciences. The method ensures that knowledge is gained through the utilisation of appropriate well-defined methods confirmed and encouraged by science theory.

Even though prudence, which is attained through experience, is said to be "useful", according to Hobbes, science, which is knowledge of consequences, is "infallible". To this famous pioneer scholar "as much experience is prudence, so is much science sapience." The use of the scientific method as tools to aid one gains knowledge is imperative as these quotations illustrates:

"The ideal of science is to achieve a systematic interconnection of facts. Isolated propositions do not constitute a science. Such propositions serve merely as an opportunity to find the logical connection between them and other propositions."[128]

"Science does not desire to obtain conviction for its propositions in *any* manner and at *any* price. Propositions must be supported by logically acceptable evidence, which must be weighed carefully and tested by the well-known cannons of necessary and probable inference. It follows that the *method* of science is more stable, and more important to men of science, than any particular result achieved by its means."[129]

"General propositions can be established only by method of repeated sampling. Consequently, the propositions, which a science puts forward for study, are either confirmed in all possible experiments or modified in accordance with the evidence. … By not claiming more certainty than the evidence warrants, scientific method succeeds in obtaining more logical certainty than any other method yet devised."[130]

"The method of science is thus essentially circular. We obtain evidence for principles by appealing to empirical material, to what is alleged to be 'fact'; and we select, analyze, and interpret empirical material on the basis of principles. In virtue of such give and take between facts and principles, everything that is dubitable falls under careful scrutiny at one time or another."[131]

Axiom 4

Let it be granted that learners who utilise the secondary law of intelligence (SLI), in addition to their employment of the blood language (BL) in imbibing knowledge, are capable of developing qualitatively well in intelligence in a higher degree that ensures the acquisition of the learner with originality and accuracy in carrying out responsibilities than those who concentrate only on the BL.

Theorem 11

Any learner who possess BL and SLI will, if examined on a particular test in connection with originality and accuracy will perform better than the individual who have not these at his disposal.

Theorem 12

---

[128] Cohen & Nagel (1949).
[129] *Ibid.*
[130] *Ibid.*
[131] *Ibid.*

The intelligence of any learner with BL and SLI will in a high degree show some originality and accuracy in performing task in the classroom context.

Theorem 13

If a learner utilises his BL and has access to SLI, his intelligence level will rise in all areas of knowledge.

Theorem 14

Any learner who uses his BL and has access to SLI will cease to be naive and his intelligence will not be displaced.

Theorem 15

If knowledge is imbibed by the learner who have access to BL and SLI, his originality and accuracy in the science subjects will rise together with his self-confidence.

Theorem 16

The work of all learners with BL and SLI will show some originality and accuracy.

Theorem 17

If a segment of population utilises the BL and has access to SLI, those that utilises BL and have access to SLI will show more originality and accuracy than those that have no access to these.

Theorem 18

If access to BL and SLI gives a learner originality and accuracy, then those that lack these will remain naive, backward and not progressive.

Theorem 19
If originality of carrying out responsibilities is increased by the use of the blood language, then access to SLI will increase it to a higher degree.

Theorem 20

If originality and accuracy are increased and perfected by the use of blood language, then additional possession of the SLI will make learners excel in mathematics and other science subjects.

Theorem 21

If learners through the use of blood language (BL) increase their capacity to analyse and describe well, then these students will show a great deal of progress in carrying out task in the classroom setting and in the society at large.

Theorem 22

If a nation uses its BL and has always access to SLI, this knowledge will make them confident, and if they had made proper alliances with successful nations, they could through adaptation process make headway for a successful development.

Theorem 23

If a nation utilises its BL and yet has additional access to SLI, it will be immersed in development concerning its vicinity, and this complete submersion into local development, will ease their great pressure, as SLI is infallible. This will yield great success to their development.

Theorem 24

If a segment of a population has neither access to BL nor SLI, it will be naive, backward, and not progressive, and they will be lagging behind all time compared to the population that has employed their BL and has access to SLI.

Theorem 25

Given what has been proved already, a nation that suddenly realises its shortcomings and make a change by employing its BL and then seriously championing the use of SLI will witness progress both at home and abroad for their development.

Theorem 26

If any nation should pay a greater attention to its use of BL and follow the principles described in primary, secondary, and main frontis while using the processes of imitation and adaptation, that nation will soon receive recognition from other nations, and its popularity will rise.

Theorem 27

The use of SLI encourages the use of imitation in the imbibing of knowledge in the academic circles; this same principle are employed in the world of development and these imitation approaches, which includes the use of BL, should take place at the imitators' own cultural context.

Theorem 28

If any nation C has a neighbouring nation F that had already been successful in their development, the nation F that had already developed will serve as a mentor that will inspire the novice nation C. Through imitation and adaptation, the novice nation C will assume a new role regarding the originality with their products, and this nation will sooner or later be reckoned with as a serious nation by the world.

Theorem 29

Let any nation choose a nucleus language as BL (spoken by the majority of its citizen, and not a foreign language), and let that nation make the central decision to employ only SLI to solve its problems. If now this particular nation gathers all its talented/geniuses among them to solve these problems at a special locality or an institution, and entrust them with these. They will be able to solve this particular problem in no a shorter period or a later date.

Theorem 30

Suppose that there is a nation that had made plans to develop; and while at the primary frontis seeks help from a wealthy developed nation. Let DA be this developed nation and SF the novice nation and let SF choose a BL and make it their sole aim to employ SLI in their development. Then, if SF decides to maintain its BL and use SLI at the same time, it will develop faster than another nation FA who neither employs its BL, SLI, nor had no DA as mentor nation or aid.

Scholium

The use of the blood language undermines the theory of superior species, as well as the theory of Lower Races. It is employment of the language of the blood that makes many stupid people look bright among the so-called developed nations. All are not intelligent as it has been believed or traditionally proposed. It is also the absence of the use of the blood language that obscures the geniuses in the developing nations and the majority of the Third World Nations. When the latter is righted or corrected, though the latter societies may not be as developed as they are now, still the differences between them and the developed nations will be marginal. The use of the blood language will see to it that knowledge will not be wasted on the learner, as the latter will shine and be able to challenge his society and the world at large.

General Scholium

This study commenced with the aim of unveiling the contribution each learning perspective has for our understanding of learning and classroom communication. A minor purpose has been to discuss how these perspectives could be blended to provide us a much useful perspective on the phenomenon of learning. Moreover, to propose additional theory of the theory of blood language. With literature from different sources and empirical investigations done in the field of educational psychology, I have tried to analyse and enlighten us on the different major perspectives of the phenomenon of learning in the classroom. It is very important for me to give credit to the articles and books that illuminate on learning and communication. In these, one clearly comprehends the central concepts which form the main ideas of each perspective. Again, the contentions between the first three perspectives, namely the sociocultural perspective, the language-focused approach, and the cognitive constructivists' perspective, are encountered. As can be seen, each perspective tries to criticise the other for neglecting this or that issue or not taken into greater consideration in this or that viewpoint. The first disagreement is about the existence of the mind and cognition. What people wonder about is whether mental processes exist as an inherited thing or whether they could be found in social interaction that occurs between the learner and the group.[132] The second conflict is put like this: supposing one agrees that cognition and mind are in fact built in the processes of social interaction, how can classroom experience which instils learning and transformation to learners be explained? It is the third contention which I should like to

---

[132] Cf also the work of Nuthall (1997).

comment on and probably offer others' views as well as my own personal view. This third conflict between these three perspectives reveals itself in the manner that existing knowledge or beliefs are transformed or restructured to generate new thinking and knowledge. The core of this contention, it seems to me, concerns the location and manner of change in knowledge and skill. The sociocultural adherents contend that change takes place while the process of acquisition is in progress. Change, they concur, ensues as a result of the interaction which takes place between a child and his parents, or between a novice and his master. Internal change, which is caused by the need for consistency and coherence between fresh experience and existing structure of knowledge, is the view of the cognitive constructivists.

The sociocultural as well as the linguistic approaches, because they stress on currency and the intrinsic quality of experience, overlooks the fact that acquired knowledge needs to be maintained throughout the individual's life span. Moreover, they are not able to make it clear how the learners' experience accords the connection that exists between different occasions and diverse settings. In contrast, the cognitive constructivists shun the major role exhibited by language, culture, and social experience in structuring the manner learners' minds function in their organisation of themselves.

These conflicts, as suggested by some researchers, will be settled if researchers would pay attention to analysis of the reception of knowledge, which takes into consideration "transactional relationship" coined by Salomon that occurs between sociocultural experience and the self-organisational mind's activities. According to Nuthall, it is the work of the neo-Piagetian that illuminates on how such an analysis will be materialised. The neo-Piagetian research deals exclusively with the manner *interactive social processes* and *structures* decide the direction of cognitive development of the child. Nuthall quotes a few researchers such as W. Doise, G. Mugny, and A. N. Perret-Clermont who have written extensively on sociocultural perspective as well as the cognitive development of the children. These scholars belong to a group of Europeans researchers who write on sociocultural perspective and think that Piaget's analysis of the development of the mind can be widened to embrace the sociocultural world of the child. In other words, in addition to social processes, social structures are necessary ingredients in the development of cognitive structures. I certainly agree with those who think that there should be a transactional relationship between these perspectives. It seems to me that any analysis of these perspectives without the inclusion of social interaction will fall short to a unique way whereby learning and internalisation of knowledge is gained. The community is as important as the child's mental processes.

Though phenomenography's approach of doing research has been recognised by the academic community, at present two major obstacles confront its wider use and credibility. These two criticisms that have been levelled against the phenomenographic approach appear in the work of Roger Säljö. The first is on how the researcher in his attempt to gather data for his research become concerned with his own manner of acquiring information and as a result depreciated the two-sided affair communication is. According to Säljö, the phenomenographic approach sometimes "uncritically allows the definition of the situation by the dominant party–the interviewer–to stand unchallenged and unproblematic." He indicates that the fixation of a shared perspective on what is talked about seems to be viewed as a problem of method only, and not as something that has to be negotiated continuously during human encounters. Secondly, while phenomenographic studies talk about learning as a change between conceptions (which sounds psychological), researchers who adhere to this method of study avoid the use of a psychological language or refrain from talking about psychological phenomena, modes of learning, and educational experiences. Säljö is puzzled concerning why they talk about a phenomenon such as learning without utilising a language that directs to events and experiences of a psychological nature. I personally think that if phenomenography approach of explaining how learning occurs in the individual will gain a wider audience, the employment of some psychological language in addition to current concepts will do well to attract more followers than it has at present. For the past twenty-five years this approach has been employed successfully to carry out research in the area of learning. I believe that to be able to penetrate into the other disciplines, both Variation and Phenomenographic approaches have to recognise its need of some psychological languages to make them attractive to behavioural scientists and the science community as a whole.

Finally, research investigations concerning the phenomenon of learning ought to concentrate on the use of the language of the blood by the learner. At the moment, it seems the individual difference as well as species intelligence differences will be bridged if all learners utilise their languages that are fixed in their blood. The focus on intelligence should be directed to this; that is, this crucial denominator, and let educators conduct intelligence tests only on learners that have used their respective blood languages in educating themselves. This will do justice to the learner and his society.

APPENDIX B

Of Darwinism and the Development of Superiority Complex Personality-type

Introduction

Ever since William James, the physiologist, philosopher, and psychologist propounded his famous theory on different personalities attracted to the domain of religion and its religious experiences, we have been honoured with additional genius enquirers who have also provided us with rich theories that concern human nature. James, an eminent philosopher of his time, for example, hanged around with many physicists such as Ernst Mach, Carl Stumpf, Hermann Helmholtz, and a few others. This was to be emulated by Sigmund Freud, a psychiatrist and neurologist, who later developed his own theories on human psyche with many dynamic terms borrowed from the physicists and natural scientists as well. Of late the Theory of Superiority Complex, which has been developed by me, used data and certain terms borrowed from the physicists and philosophers domains just as James and Freud had done that makes the theory to be classified as lying in the domain of psychophysics. This present work emerges from the constant and meticulous observation made by me in connection with human nature, and the used of historical data accumulated over the years concerning man that has been gathered and analysed by eminent philosophers and physicists.

The Purpose

The purpose of the present chapter is to analyse and make it more available that the same knowledge that helped removed the primitive man from his depth of poverty and ignorance (so that he could later enjoy contemporary lifestyle that modern technology has brought) has had with it its negative repercussions. In other words, what enabled man to succeed in his adaptation to this planet earth also helped set a trap for the downfall of man (at least with regards to the brutal behaviour of dealing with one another on this planet) in the successive generations. The central thesis of this chapter is that Darwinism, which through laborious effort unveiled the origin of the modern man, generated a condition that set the modern man on to his kneels again. This paper will argue that the idea of setting certain races or giving hierarchy in races, one above the other and denoting some as superior species sparked off the gradual development of a psychological condition of human illness, which resulted what I have termed a "psychological superiority complex". Such a miserable condition, which we hardly recognise, its durable costs and refused to give it a psychological analysis has been hunting men the last hundred years and throughout our technological ages. How the global condition as a whole made it possible or contributed to the acceleration of this dangerous doctrine that had left its deadly mark on the planet earth will be probed. But before I come to this examination, I shall review the central elements of Darwin's theory, and then proceed to the analysis on how the nineteenth century scholarly community reacted. It is an accepted fact that our noble Carl von Linné, the Swedish botanist whose work preceded that of Darwin, was on the threshold of propounding this important theory.[133] Then finally, I shall propose what characterises people who have been attracted to this so-called theory of superior species that I have elsewhere disproved as a false notion with elaborate elucidation.

Basic Scientific Ideas Preceding Charles Darwin

Scholars as well as most thinkers generally agree that what Newtonianism had been to the eighteenth century Darwinism was to be to the nineteenth. It was the pivotal point in a crisis of faith and in a whole series of intellectual revaluations in philosophy and social thought. Darwin's theory animated scientists seeking truth and imperialists seeking colonies. Prior to Darwin's publication of *The Origin of Species* in 1859, rationalist criticism had struck hard and wounding blows at Christianity. The rationalists undermined not only its teachings of revelation and the miraculous but also the conception that its morality gave the sole sound basis for human guidance. Speculative thinkers had already made available the belief in changing species and in evolutionary development to many educated men as far back as the time of Aristotle, and also right from the previous century before the publication of Darwin's book. Therefore, the case of evolutionary thinking was not new. Many important names and unforgettable thinkers such as Buffon, Lamarch, and Charles Darwin's own grandfather, Erasmus Darwin had prepared the minds of people in the first half of the nineteenth century on the issue that the world was much older than it appeared in Biblical chronology. Even the book by C. Lyell, *Principles of Geology* published in 1830, had made it possible the unveiling of the theory of uniformitarianism which states that the geologic history of the earth had been slow, uniform, orderly development brought about by the accumulation of constant small changes. Lyell's theory, according Garraty & Gay, was "of course, profoundly at odds not only with the prevailing view among scientists but with the Biblical version of the creation and the flood." More so "cautious as Lyell was about the theological implications of his own

---

[133] Nilsson (2007:26-31). Carl von Linné's greatest contributions were his books Systema Naturae and Sexual Systems which he published while he was a professor at the Uppsala University.

findings, his authority weighed heavily toward a new view of the antiquity and history of the earth that was consistent with the evolutionary idea" [134] and that was Lyell's advantage.

As some well-known thinkers have constantly stressed Darwin's argument is easily misrepresented. The most overlooked element in his ingenious theory of evolution, according to Maddox, is that it is not a theory about individuals, but about groups of individuals belonging to the same species, that is, a population.

Let us take for instance a population of animals or plants in which sexual reproduction is the norm and consider also the milieu in which these creatures are placed. There may be other species living in this environment, which will certainly compete with all members of the population for common resources such as food, and may even be predators of them. The progeny of their matings will generally correlate their parents, but will also by chance externalise small variations from the norm. According to Darwin argument, these variations are heritable. This being Darwin's first pillar of his theory is regarded as chance but heritable variation.

According to Darwin, many variations will be disadvantageous for the individuals concerned (they may die or fail to reproduce), but some will be advantageous in a particular environment at a particular time. The advantageous variations give the progeny carrying them a great chance than others of producing offspring of their own; these characteristics will therefore become more common in succeeding generation. That is natural selection. As John Maddox (1998) comments, "although the process hangs on the survival and fecundity of individuals, what matters for evolution is contribution of individuals' progeny to the next generation." "As generations go by," he continues "some characteristics may become more and more common, perhaps conspicuously so. Natural selection, which moulds this process of adaptation, is the second cornerstone of Darwinism."[135]

The Central Elements of Darwin's Theory

To present the basic elements of Darwin's theory of natural selection I would make use of an unpublished paper of 1848. This was eleven years prior to the publication of *The Origin of Species* which was also utilised by John A. Garraty and Peter Gay *The Columbia History of the World* published in 1984 2ed. As Malthus and Spencer had done before him, Darwin through his meticulous observation found that *firstly*, within each species more organisms are constantly generated that can be nourished and supported by the environment. *Secondly*, the rapid rate at which living organism or forms increase produces a constant struggle for existence, a competition for food and other means of survival. *Thirdly*, some variation of physical type always occurs within a species. In short not all organisms are adequately and equally equipped for survival: those whose variations are better adapted to the environment in which they must live are the ones that survive and reproduce themselves. *Fourthly*, the offspring of these survivors inherit their variations. *Finally*, the accumulation of these small favourable variations over a very long period of time introduces such changes of type that a new species emerges. On the strength of Darwin's theory, and of an examination of the distribution of animal forms on the face of the planet earth, one could draw up a hypothetical picture of the development of complex forms from simple ones, a picture of evolutionary change commencing with rudimentary forms and ending with man. But these central proposals were not to create deep worries until in 1871, where with the publication of *The Descent of Man*, Darwin made it strongly clear that his theory embraced human origins as

---

[134] Garraty & Gay (1984:955).
[135] Maddox (1998:237).

well as others. This plain declaration upset many people who had been to consider the idea of changes in all the other species.

Social Darwinism: The Four Phases

We know from history that even among the so-called Christians Darwin's found supporters and it would have been a greater influence had it not conflicted with creation and some major fundamental beliefs of the Christian Church. Yet the theory found it way among many people who translated Darwin's main hypotheses into four social categories. In the first phase, the struggle for existence became economic competition or war; in the second phase, survival became economic or military predominance; in the third phase, the idea of inheritance of variations was further evidence of necessary human inequality; and finally, in the fourth phase, adaptation to environment was elevated to a social as well as a biological value. In the words of two prominent historians, many philosophers or thinkers considered the whole scheme of development to be a promising analogue of social progress. Indeed there were other thinkers who emphasised that the mechanisms of natural selection were the chief means, probably the only means, of obtaining that progress. The conception of geological periods of time came to serve as a model for society. That implies long periods of time would be necessary to bring about any social change, as argued by conservative theorists, just as it had taken aeons to produce the highest forms of animal life. Therefore, all schemes for rapid reform fly in the face of nature.

Though the translation of Darwin's hypotheses into social categories found numerous critics and opponents, as well as those who wondered whether these were appropriate and could they be able to construct analogies between society and nature (i.e., life of men in society as a constant struggle), these became intensified when Karl Marx took these interpretations further. According to Marx, Darwinism should not be restricted to individualists and the nationalists; the real struggle, Marx indicated in his writings was between classes. It can be asserted that Darwinism provided Marx a basis in natural science for the class struggle in history. Later, social scientists joined in these argumentation and themselves concluded that the important struggle to consider was not among nations, classes, or individuals, but among institutions, habits, and types of character, though this took them far away from Darwin's thinking.[136] Later during the end of the century Darwinism was put to two mutually antagonistic uses at its lowest level. In the first place, Herbert Spencer in England and W. G. Sumner in the USA put it into the service of conservatism, industrial capitalism and laissez-faire. Sumner, who in 1880 laid down the essential tenet of this point of view, argued that "the millionaires are a product of natural selection, acting on the whole body of men to pick out those who can meet the requirement of certain work to be done."[137] But later in 1893 Sumner regretted having asserted that when he realised that the hopes of the Social Darwinian philosophers had not been justified. In his words we read, "The doctrine of evolution has not furnished guidance to the extent I had hoped. Most of the conclusions, drawn empirically, are such as right feelings, enlightened by cultivated intelligence, have already sufficed to establish."[138]

The racist-militarist, which characterises the second phase of Social Darwinisn, was particularly influential on the continent of Europe, though it was visible also in the American society. It was with regards to this that in 1872 the English economist, Walter Bagehot concluded in his *Physics and Politics* that those nations, which are strongest, tend to

---

[136] Garraty & Gay (1984:959).
[137] *Ibid.*, p. 956.
[138] *Ibid.*

prevail and that in certain marked peculiarities the strongest are the best. In Europe, some well-known thinkers elevated the Social Darwinian ideology to a much more penetrating militarist doctrine. Joining religion with evolutionary science the German Marshal Helmuth von Moltke advocated that "War is an element of the order of the world established by God. Without war the world would stagnate and lose itself in materialism."[139] Even before the terrible bloodletting the World War One caused, most philosophers as well as some laymen repudiated this dangerous ideology. It also had no support whatsoever among serious thinkers any part of the inhabited planet earth.

But there was, however, a rebate of Social Darwinism in the ideologies of the Fascists, stated both by Mussolini and by Herr Adolf Hitler. According to Hitler, "the stronger has to rule." "Only the born weakling can consider this as cruel. The fight for daily bread makes all those succumb who are weak, sickly, and less determined." With regards to this quotation, Garraty & Gay (1984) comments is worth noting, "With its harsh militarism and its genocidal mania, twentieth-century totalitarianism seemed like a ghastly caricature of the relatively innocent ideas of the nineteenth-century Social Darwinisms, and its history suggests that this form of Social Darwinism could recur as a serious strain in human thought only with a grave lapse of humankind into barbarism."[140]

Criticisms from Modern Psychiatry Point of View

Darwin and many of his contemporaries were right as regards evolution and its associated characteristics that occurred with certain species that helped them in their adaptation. Even today we see natural selection as functioning in the gradual evolution process in plants as well as in the animal world.[141] Those that are capable of surviving from the dismal of the planet earth are able to produce and function well in their milieu. But it was also noted in Darwin's view that "even in a species in equilibrium with a stable environment (or in stasis), new variations will continually arise at random, but will not usually change the general character of the species."[142] But let us take for instance that a population of some species has reached equilibrium in some stable environment, and that the ambient temperature then begins to increase. According to evolutionary theory "The members of the population carrying the alleles best suited to higher temperatures will prosper in comparison with their fellows, so that those alleles will be more common in the next and later generations."[143] This means "mutations that might previously have sharply reduced the fitness of the organism may also appear and be perpetuated. The consequence will be a changed population, with different frequencies of the old alleles and perhaps novel alleles as well." [144] Another hint, which is mentioned by Maddox, is the continuing ignorance about the differences between a species and the ancestor species from which it has evolved. There is straightforward operational distinction between living species. Members of distinct species are mutually infertile. It was Darwin's meticulous observation of the finches of the Galapagos Islands that led him to draw the conclusion that prolonged geographical isolation is one cardinal route to the emergence of new species. It appears that two identical populations may drift so far apart that their members can no longer interbreed. Darwin made it available in later writings about what he called sexual selection, "dependent on supposed ingrained mating preferences, which he recognized

---

[139] *Ibid.,p.960.*
[140] *Ibid.*
[141] Maddox (1998).
[142] *Ibid.*
[143] *Ibid.*
[144] *Ibid.,pp.247-248.*

to be another means by which a population might separate into at least two parts that are reproductively isolated from each other." As time elapsed "subgroups of the population would then be affected differently by natural selection. Darwin supposed that the passage of generations would yield mutual infertility and new species." According to Maddox, "what was missing, and is still missing, is a rounded understanding of the genetic causes of mutual infertility" and he believes an understanding of this mechanism of inheritance is in the area of genetics.[145]

But what one wonders is whether there is something inherently unique about specific novel species in this evolutionary process or development that can make us attribute "superiority" to a particular species. Should there be any hierarchy in the human family tree as suggested by pioneering scholars who propounded this theory so many centuries ago? Why should we ever consider this development as following the vertical development (from the lower up to the upper level) one and not the horizontal one (that shows minor specific differences)? Many years ago these things regarding superior (adaptable) qualities could be accepted without much trouble or thinking because we were so much ignorant and biased about some issues that pertain to the human being and his gradual evolving of his adaptable qualities. Surely adaptation is a hazardous business. Today science informs us about many things and reveals secrets that we were not capable to explaining in those centuries where religion dominated mankind and everything was subjected under the authority of religion. The whole educational enterprise was one-sided, only specific groups of people were mainly concerned with research and theorising. We have through meticulous observation gained the knowledge that though adaptation may be impelled by several influences, the environment of the living organisms determines to what extent these organisms are capable of mustering all mental resources and energy to help them in their adaptation. Science, the philosophy of the enlightenment era, has furnished us that mankind made a great deal of progress when they submerged their mental faculties to the gaining of scientific knowledge and the implementation of scientific methods to help in the adaptation. Advancement of knowledge which gain currency in these periods and the utilisation of the "blood language" to imbibe knowledge helped the human organism more than any so-called inherent qualities rather gained from natural selection and evolution itself. Analysis show that there must be a possible explanation as to why others have since the last 100 years gone ahead with much development that make them consider themselves fortunate or regard themselves as "superior". That is, if we were to disregard certain myths about human evolution, then we must through rigorous methods find a suitable explanation why these things have taken place, using sciences instead of the favourite accumulated myths particularly used by a segment of the human race to extol themselves. To me the use of psychophysics and chemistry and not only *biology* would be capable of providing us better answers as to why certain humans have been fortunate to surge ahead in development which has taken place the last hundred years, and for some only sixty years ago. It is therefore appropriate to point out that my proposition that "Superiority is attained but not given" should be made the object of examination in connection with the theory of evolution. In this case, Darwin's theory is not threatened nor rejected. Rather my "Theory of Superiority Complex" which utilises this proposition to predict how every species of the human race has the best possibility of reaching to the pinnacle of development is sound and put Darwin's theory in a rather better perspective.

Darwinism and the Psycho- S Complex Personality-type

---

[145] *Ibid.*

One does not have to strain his mind in order to recognise the implication of Darwin's theory concerning the so-called new species or novel species that finally emerges. In fact, experiments with fruit-fly species and a few documented cases have shown that point mutations in some gene, the simplest form of genetic change, have led to separate species. But this is not important for us now at this point of our analysis. Though the suggestion of a human new species was not mentioned to signify an "extraordinary one" sadly enough that was the implication ascertained from Darwin's theory by the majority of people both laymen and the scientists who became conversant with his theory. Therefore, it was not out of place when the theory eventually bolstered up the spirits of scientists looking for truth regarding everything concerning nature, and imperialists then seeking for colonies in their quest for domination which was predominant in those eras of human history. One of the laws of the Theory of Superiority Complex propounded by me states that "Superiority complex is in human beings universally, but its intensity is proportional to the increase gained in knowledge". A person may think of himself as higher than his subordinate, his employee, another person from the different part of the world, or think for example about the caste system in India. In this case, our analysis of superiority complex is geared toward diverse groups of people throughout the whole world (known part of the inhabited universe) and not to a specific group of people or race.

In the "Theory of Superiority Complex: Personal and Collective Experience" propounded in the six chapter of this book I used the scientific deductive system in developing it. In this chapter, I clearly defined what I meant by "superiority complex" as indicating a brooding over the idea that one is a superior being or higher than others. This concerns looking down to others, as they do not measure up to the standard of human beings as oneself; in other words, they are human animals (*urmänniskor*). This definition was connected to "fear" or "a threat" that the individual with superiority complex faces or imagines in his head concerning those he tries to subject or subdues as human animals. The word "complex" as a psychological term encompasses a related group of usually repressed feelings or thought which cause abnormal behaviour or mental states. The word denotes a preoccupation or an obsession with something, which in this context is "superiority" or a "better species" than other species. What I did not do in my previous work was the lack of readiness to present the *psychological states* (psychiatric symptoms) of those people or individuals that become ensnared into this miserable popular condition.

Psycho-S Complex: Signs or Symptoms

Now I shall analyse briefly the "signs" or symptoms that characterise the individuals who through superiority complex disorder become obsessed with hostilities toward others.

Delusion

A Psycho S Complex individual greatest suffering is that he suffers unprecedented delusion. The sort of delusion he undergoes makes him sees the letter A as B or C even though all the people standing around visualise the first-mentioned letter. There is nothing seen as deliberately imagined or conscious regarding these matters. The Psycho S Complex person's world is filled with human animals of different kinds and colours that do not measure up as he and his human beings' world of people do. The delusion will make him/her see, for example, the Jews as not human beings as they are even though history has shown us that the Jews have produced the most intelligent people on earth that have presented many scholars who have been awarded the Nobel Prize. The delusion is usually reported as having the complexion of

aggressive behaviour and *barbarism* which characterises his manner of persecuting *firstly*, himself and *secondly*, others that he regards as human animals.

Auditory hallucination

A Psycho S Complex individual is constantly plagued by auditory hallucinations as it happened to Saul of Tarsus and Saul, the King of Israel in the Biblical mythology and also Henrich Himmler who persecuted many innocent Jews during the Second World War. It was said that the Psycho S. Complex King Nebuchadnezzar of Babylon suffered auditory hallucination that made him killed the Jews and many neighbouring tribes in the Ancient Empire. His madness even compelled him to go and sleep in the wilderness for a period and ate leaves as his main food. There are proofs that the Psycho S Complex attaches great importance to his ***dreams*** as he usually set out to fulfil them to the letter. Many Psycho S Complex individuals attached great importance to the dream or prophesy fabrication "King" Nostrademus's works during the Second World War in Europe.

Prejudice predisposition

A Psycho S Complex individual is indescribably bias in his crooked small world that are filled with both animal human beings and real human beings. He does not see anything good in other species except his own that he has been taught to believe that he is of a "superior origin". This miserable disposition makes it very difficult for him to entertain others who are not of his kind (human beings) in his vicinity. This bias predisposition explains the reason why even in terms of job application he would not recognise the qualification that comes from the same country that presents all the merits needed for a specific job in question. She refuses to give the job out not because a person does not possess the qualification or there is someone better than this person is, but because he is not of a "superior" species or origin. His illness, which is *narcissistic*, will allow her to recognise only the person who looks like her as a human being. All other persons are regarded as human animals (*urmänniskor*).

Manipulative

This word comes from Latin *manipulatio*, which is influenced by the French word *manipuler*. It means that to manage a situation or a person to one's own advantage, especially unfairly or unscrupulously. A Psycho S Complex individual greatest weapon that in his delusion makes him the cleverest among all that he looks down upon as human animals is that he knows how to manipulate other people into believing that he is "superior" or whatever he claims to be. Manipulative tactic is one of the tactics that are enshrined in his propaganda weapons, and they are mirrored in his delusions which make him hold on to that "illusory superior power" he claims to have among the numerous people he misleads. Among other things, there are so many secret tactics that through his strenuous campaign to dominate the individuals around or the world he lives in he employs but these are not accessible to these innocent and kind-hearted individuals. It is said that when a person is hungry he will do everything that is expected by his captors to satisfy his hunger or quench his thirst. The manipulative Psycho S Complex is aware of this and therefore employs it frequently. Manipulative tactics can be used even to cause hunger and then be able to secure the obedience of the poor hungry persons who trust him. By providing these needs, which is a sort of camouflage, he will pretend that he is the kindest and most generous of all individuals on earth.

Winmust

This is to denote the Psycho S Complex individual's inclination to win in all things or challenges that confronts him/her. In his delusion he always sees himself as a winner even when he is loosing or had already lost. His imagination of being superior does not allow him to accept reality as it is when cornered or faced with difficulty. The *winmust* usually compel a Psycho S Complex to face a situation where he falls into the *dissonanspositioonis*. This is defined as *a state where a person finds it uncomfortable to admit a situation because of its unfavourable character*. We see this taking place in the Second World War when Herr Adolf Hitler through propaganda assured the German people that the Russians had been driven back to Moscow, whilst meanwhile the Russians had already crossed the border and were already on the verge of entering Berlin.

Superexaggerare

A Psycho S Complex individual delights in exaggerations because of his desire to dominate or subdue all others below unto his feet. He doubles the magnitude of certain incidents. Supposing that through environmental pressure and difficulty he is taxed to work hard to overcome his immediate problems and adaptation he will easily attribute his success to his being of a "superior species" or his "genetic make-up" being different from all others. Meanwhile, there are many indications that say otherwise that it was through poverty, hard work, and frustration that made them succeed. Exaggeration will be seen in all his activities, advertisements, and propaganda tactics so as to devalue those he considers as human animals.

Superwunsken

A Psycho S Complex individual has *superwunsken* or portentous wishes. Thanks to the sciences their fantasies and wishes are mostly attained through hard work and labour. They regret being in this world with others who through their delusion consider them as **human animals (*urmänniskor*)**. So they imagine themselves coming from other planets or others being in the category of human animals and they being human beings. They are the Homo sapiens or they descended from angels in the heavens. But, meanwhile the opposite is the true story in that those considered as human animals might have also won many Nobel Prizes (through developing theories) and have even invented many important machines or made significant discoveries. They might have helped these "human individuals" to have been able to develop their own environment and also these human beings own language. They may have produced important scholars as well as many geniuses in the world and probably one of them was recognised as the most intelligent person on earth. The human animals make one of the most intelligent groups in the world. But Psycho S Complex through his delusion sees otherwise, because of his crooked worldview or obsession with superiority or his *winmust* to dominate the poor and the weak.

Infantile notions

The contents of the ideology a Psycho S Complex holds are filled with "omnipotent" wishes and infantile notions, such as, "superiority," "I am the best," "I want the greater share," "we are the masters of the world," "they are human animals and we are human beings" or "Homo sapiens". Since the ideology does not grow in comparison with his own mature faculty and influence, *he is constantly troubled and this kind of influence* is inversely revealed in war and aggression to his neighbours or those he rules. If we were to allow the Psycho S Complex individual alone so as to obtain his wishes he would go to war for several centuries since he delights in *domination* and *aggression*. Those he succeeds in winning their influences then behave like robots and could accept these infantile notions without applying the critical faculty to examine these childish contents. This makes them susceptible to criminal tendencies such as *sabotages, conspiracies,* and **secret usurpation of foreign governments,** and *aggression.* In fact, many of them eventually earn up becoming criminals and sent to prisons or *rule* others. These things strip them their rightful respect which would have been accorded to them had there been proper influence they maintain or show in this world.

Hannibal Odyssey

War is a symptom of a disorder; in fact, it is a ***primitive manner*** of settling cases or a dispute. But to the Psycho S Complex individual, war is seen as a sign of greatness. He yearns for war. He prays for it. He rejoices in going to war. He has ***social and emotional attachment*** to war (as it is dramatises in his milieu now and then) and there is higher tolerance of aggressive impulses among the collective group he belongs. He delights in going to war and causing instability in the whole world and makes sure that adequate preparation is made long beforehand should in case there is this illness or suffers from this disorder.

Scholium

At last we have proceeded to the end of the analysis which has considered the theory of Darwin as generating a psychological disorder in this modern world. It is the objective of scientific investigation that new insights being generated should be added to the already accumulated information for them to be probed and then be accorded a place in the scientific literature. What we have done here is the proposition of an extraordinary psychological illness to be termed "Psycho Superiority Complex (psycho s complex)" which is a disorder that has eluded many eminent theorists in psychiatry as well as psychology for over a century. The importance of this article is that it has made it available the *symptoms* that should serve us as a guideline for the diagnosis of this disorder in our modern society. It is the intention of the originator of this theory that more consistent analyses and investigations will yield better results that may serve the public who we, as responsible researchers, owe our responsibilities. Let the scientific community make this disorder their object of research! Let them accord it a place in the psychiatry literature and be spread out in scientific journals as well!

Inevitably with a theory as new, as unusual, and as conforming to genuine scientific method of deductive, dissension will soon emerge as regards its acceptability in the academic world. But such a conflict should not distract us; the theory of superiority complex should be recognised and be given a place in the scientific literature and journals. The theory's great predictive power should make us give it a more thorough attention and be accepted as one of the most useful theories in the twenty-first century.

List of The Principal And Minor Works Consulted.

Achtenhagen, F. (1997). Teaching across the curriculum areas. In: *An Evaluation of Swedish Research in Education.* (Eds.) Rosengren, K. E. & B. Öhngren. pp. 106-119. Uppsala: Swedish Science Press.

Alfvén, H. (1978). How should we approach cosmology? In *Problems of Physics and Evolution of the Universe.* Academy of Sciences of Armenian SSR, Yarevan.

Alfvén, H. (1983). On hierarchical cosmology. *Astrophysics and Space Science*, vol. 89:313-324.

Alfvén, H. (1988). Cosmology in the plasma universe. *Laser and Particle Beams*, vol. 16:389-98.

Anokhin, P. K. (1958). The role of the orienting-investigatory reaction in the formation of the conditioned reflex. In L. G. Voronin *et al.* (eds.) The Orienting Reflex and Exploratory Behavior. *Academic Pedagogic Science*, Moscow.

Ausubel, D. P. et. al. (1968). *Educational Psychology. A Cognitive View.* New York: Holt, Rinehart and Winston.

Ayim-Aboagye, D. (200). *Prison, Punishment and the Church: A Socio-Psychological Investigation of the work of the Chaplains among the Immigrants Inmates in Swedish Prisons.* Religionspsychologiska skrifter nr.8. Uppsala: Uppsala University.

Ayim-Aboayye, D. (2002). *Swedish Educational Reforms in Retrospect: The Impact of Educational Research on Policy Formation.* (In review) Oxford Review of Education

Ayim-Aboagye, D. (2003) Adult education, knowledge expansion and local steering: An analysis of some themes in Swedish education system. *(In review). Oxford Review of Education.*

Ayim-Aboagye, D. (2003) *Teaching, ethnicity and university education: Exchange Students experiences of school and learning in Sweden.* Research Proposal. Department of Education, Uppsala University.

Ayim-Aboagye, D. (2006). *Indigenous Psychiatry. Transcultural Study of Traditional Practitioners in the West African Communities.* Religionsvetenskapliga skrifter no. 66. Åbo: Åbo Akademi University.

Bandura, A. (1986). *Social Foundations of Thought and Action. A Social Cognitive Theory.* New Jersey: Prentice-Hall, Inc., Englewood Cliffs.

Bergson, H. (1976). *Les Deux Sources de la moraleet de la religion.* [**The Two Sources of Morality and Religion**] Paris: Presses Universitaires de France.

Bernstein, J. (1996). *A Theory for Everything*. New York: Spriner Verlag.

Berlyne, D. E. (1960). *Conflict Arousal and Curiosity*. McGraw-Hill Series in Psychology. New York.

Bohm, D. (1952). *Quantum Theory*. New York: Prentice-Hall, Inc.

Bowden, J. & F. Marton (1998). *The University of Learning. Beyond Quality and Competence inHhigher Education*. Kogan Page.

Bruce, A. & Wallace, D. (1989). Critical point phenomenon: universal physics at large length scales. In *The New Physics*. Pp. 236-267. (Ed.) Davies, P. Cambridge: Cambridge University Press.

Carnap, R. (1951). *Logical Foundations of Probability*. London: Routledge and Kegan Paul.

Cohen, M. R. & Nagel, E. (1949). *An Introduction to Logic and Scientific Method*. London: Routledge & Kegan Paul Limited.

Close, F. (1989). The quark structure of matter. In *The New Physics*. Pp. 396-424. (Ed.) Davies, P. Cambridge: Cambridge University Press.

Darwin, C. (1889). *On the Origin of Species...* (Authorised edition from English ed. II); New York.

Davies, P. (Ed.) (1989). *The New Physics*. Cambridge: Cambridge University Press.

Davies, P. (1989). The new physics: a synthesis. In *The New Physics*. Pp. 1-33. (Ed.) Davies, P. Cambridge: Cambridge University Press.

Dolin, A. O., Zborovskaia, I. I. and Zamakhover, S. K. (1958). On the characteristics of the role of the orienting-investigatory reflex in conditioned-reflex activity. In L. G. Voronin *et al.* The Orienting Reflex and Exploratory Behavior (eds.) *Academic. Pedagogic. Science*, Moscow.

Epstein, L. C. (1992). *Relativity Visualized*. San Francisco: Insight Press.

Ford, J. (1989). What is chaos, that we should be mindful of it? In *The New Physics*. Pp. 348-371. (Ed.) Davies, P. Cambridge: Cambridge University Press.

Gay, P. (Ed.) (1973). *The Enlightenment. A Comprehensive Anthology*. New York: Simon and Schuster.

Garraty, J., & Gay, P. (Eds.) (1972). *The Columbia History of the World*. New York: Harper & Row.

Georgi, H. M. (1989). Grand unified theories. In *The New Physics*. Pp. 425-445. (Ed.) Davies, P. Cambridge: Cambridge University Press.

Gould, S. J. (1989). *Wonderful Life: The Burgess Shale and the Nature of History*. New York: W. W. Norton.

Guerlac, H. (1961). *Lavoisier—the Crucial Year*. Ithaca, New York.

Guth, A. & Steinhardt, P. (1989). The inflationary universe. In *The New Physics*. Pp. 34-60. (Ed.) Davies, P. Cambridge: Cambridge University Press.

Harlow, H. F. and Zimmermann, R. R. (1958). The development of affectional responses in infant monkeys. *Proc. Amer. Phil. Science.*, 102, 501-509.

Hawking, S. (1989). The edge of spacetime. In *The New Physics*. Pp. 61-69. (Ed.) Davies, P. Cambridge: Cambridge University Press.

Hawking, S. (2001). *The Universe in a Nutshell. The Inspiring Sequel to A Brief History of Time*. London: Transworld Publishers.

Heath, T. L. (1897). *The Works of Archimedes*. Cambridge: University Press.

Heath, T. L. (1908). *The Thirteen Books of Euclid's Elements Translated from the Text of Heiberg with Introduction and Commentary*. 2 Ed. Cambridge: University Press.

Heath, T. L (1949). *Mathematics in Aristotle*. Oxford: Clarendon Press.

Hinde, R. A. (1954). Factors governing the changes in strength of a partially inborn response as shown by the mobbing behaviour of the chaffinch *Fringilla Coelebs*. I and II Proc. Roy. Soc., 142, 306-331, 331-358.

Hobbes, T. (1665). *Leviathan* Cambridge: Cambridge University Press.

Hume, D. ().An Enquiry Concerning Human Understanding, in Works, IV

Hutchins, R. M. (Ed.) (1952). Newton & Huygens 34. *Great Books of the Western World*. Chicago: Encyclopaedia Britannica, Inc.

Isham, C. (1989). Quantum gravity. In *The New Physics*. Pp. 70-93. (Ed.) Davies, P. Cambridge: Cambridge University Press.

Jönsson, A-M. (2007). "Ett öppet krig—per correspondens,"*Forskning & Framsteg*. Nr 1, Jan-Feb, pp. 22-25.

Kaplan, M. (1957). *Systems and Process in International Politics*. New York: John Wiley and Sons.

Knight, P. (1989). Quantum optics. In *The New Physics*. Pp. 289-315. (Ed.) Davies, P. Cambridge: Cambridge University Press.

Kuhn, T. S. (1970). *The Structure of Scientific Revolution*. 2nd Edition. The International Encyclopedia of Unified Science. Vol2., No. 2. Chicago: The University of Chicago Press.

Lave, L. & Wenger, E. (1991). *Situated learning. Legitimate Peripheral Participation.* Cambridge: Cambridge University Press.

Lave, L. (1988). *Cognition in practice. Mind, mathematics and Culture in Everyday Life.* Cambridge: Cambridge University Press.

Leggert, A. (1989). Low temperature physics, superconductivity and superfluidity. In *The New Physics*. Pp. 268-287. (Ed.) Davies, P. Cambridge: Cambridge University Press.

Lerner, E. J. (1991).*The Big Bang Never Happened. A Startling Refutation of the Dominant Theory of the Origin of the Universe.* New York: Simon & Schuster.

Lisina, M. I. (1958). The role of orientation in the conversion of involuntary reactions. In L. G. Voronin et al. (eds.) The Orienting Reflex and Exploratory Behavior. *Academic Pedagogic Science*, Moscow.

Lloyd, E. (1980). *Handbook of Applicable Mathematics Vol. II Probability.* New York: John Wiley & Sons.

Longair, M. (1989). The new astrophysics. In *The New Physics*. Pp. 94-208. (Ed.) Davies, P. Cambridge: Cambridge University Press.

Lorentz, H. A. (1931). *Lectures on Theoretical Physics.* Delivered at the University of Leiden. Vol. III Transl. by L. Silberstein & Trivelli. London: Macmillan and Co., LTD.

Maddox, J. (1998). *What Remains to be Discovered.* London: Macmillan.

Mach, E. (1976). *Knowledge and Error.* Dordrecht: D. Reidel Publishing Company.

Marler, P. (1956). "Behaviour of the Chaffinch, *Fringilla Coelebs*, Brill, Leiden.

Marton, F. et. al. (1993). Conceptions of learning. *International Journal of Educational Research*, 19, 277-300.

Marton, F. (2000). The practice of learning. *Nordisk Pedagogik*, 20 (4), 230-236.

Marton, F. (1981). Phenomenography. Describing conceptions of the world around us. *Instructional Science*, 10, 177-200.

Nagel, E. (1961). *The Structure of Science. Problems in the Logic of Scientific Explanation.* New York: Harcourt, Brace & World, Inc.

Moscovici, S. & Doise, W. (1994). Conflict & Consensus. A General Theory of Collective Decisions. London: Sage Publications.

Newton, I. (1999). *The Principia. Mathematical Principles of Natural Philosophy.* Transl. By Cohen, Bernard & Whitman, Anne. Berkeley: University of California Press.

Nicholson, M. (1970). *Conflict Analysis.* London: The English University Press Limited.
Riker, W. H. 1962 *The Theory of Political Coalitions.* New Haven: Yale University.

Nicolis, G. (1989). Physics of far-from-equilibrium systems and self-organisation. In *The New Physics*. Pp. 316-347. (Ed.) Davies, P. Cambridge: Cambridge University Press.

Nilsson, E. (2007). "Blommar som kunde gjort Linné till Darwin," *Forskning & Framsteg*. Nr 1, Jan-Feb, pp. 26-31.

Nuthall, G. (1997). Understanding student thinking and learning in classrooms. In B. J. Biddle, T. L. Good & I. F. Goodson (Eds.) *International Handbook of Teachers and Teaching*. Vol.II (pp.681-768). Boston: Kluwer.

Partington, J. R. (1951). *A Short History of Chemistry*. London.

Permutter, M. & Hall, H. (1990). *Adult Development and Aging*. New York: John Wiley & Sons.

Piaget, J. & Inhelder, B. (1969). *The Psychology of the Child*, Basic Books, Inc.

Piaget, J. (1972). *The Principle of Genetic Epistemology*. Transl.: Mays, W. London: Routledge & Kegan Paul.

Piaget, J. (1972). *Psykologi och Undervisning*. Stockholm: Aldus Bonnier.

Piaget, J. (1974). *Sociologiska Förklaringar*. Lund: Studentlitteratur.

Prigogine, I. & Stengers, I. (1984). *Order Out of Chaos*. New York: Bantam Book.

Rogoff, B. (1990). *Apprenticeship in Thinking*. Oxford: Oxford United Press.

Rosengren, K. E. & B. Öhngren (eds.) (1997). *An Evaluation of Swedish Research in Education*. Uppsala: Swedish Science Press.

Ross, G. R. T. & Haldane, E. S. (Transl.) (1979). *The Philosophical Works of Descartes*. Vol. 2. Cambridge: Cambridge University Press.

Runesson, U. (1999). *Variations Pedagogik. Skilda sätt att Behandla ett Matematiskt Innehåll*. Göteborg Studies in Educational Sciences 129. Acta Universitatis Gothoburgensis.

Råde, L. & Westergren, B. (1988). *Mathematics Handbook. Concepts, Theorems, Methods, Algorithms, Formulas, Graphs, Tables*. Studentlitteratur. Chartwell-Bratt LTD.

Salam, A. (1989). Overview of particle physics. In *The New Physics*. Pp. 481-491. (Ed.) Davies, P. Cambridge: Cambridge University Press.

Schilpp, P. A. (Ed.) (1949). *Albert Einstein: Philosopher-Scientist*. Evanston, Ill.

Schmitz, H. (2007) "Linnés passion," *Forskning & Framsteg*. Nr 1, Jan-Feb, pp. 18-21.

Shimony, A. (1989). Conceptual foundations of quantum mechanics. In *The New Physics*. Pp.373-395. (Ed.) Davies, P. Cambridge: Cambridge University Press.

Skinner, B. F. (1953). *Science and Human Behaviour*. New York: Macmillan.

Skinner, B. F. (1974). *About Behaviourism*. New York: Knopf.

Skinner, B. F. (1983). Intellectual self-management in old age. *American Psychologist*, 38, 239-244.

Smith, N. K. (Transl.) (1982). *Immanuel Kant's Critique of Pure Reason*. Kant, E. (first published 1929). London: Macmillan Press LTD.

Suppes, P. (1957). *Introduction to Logic*. New York: D. Van Nostrand Company.

Säljö, R. (1996). Mind action- conceiving of the world versus participating in cultural practices. In: *Reflections on Phenomenography. Toward A Methodology?* (Eds) Dall'Alba & Björn Hasselgren. Göteborg Studies in Educational Sciences 109. pp.19-33.

Säljö, R. (2000). *Lärande i Praktiken*. Stockholm: Prisma.

Taylor, J. (1989). Gauge theories in particle physics. In *The New Physics*. Pp. 458-480. (Ed.) Davies, P. Cambridge: Cambridge University Press.

Thompson, D. (1995). *Oxford University Dictionary. 9th Edition. The foremost authority on current English*. Oxford: Clarendon Press.

Thom, P. (1981). *The Syllogism*. München: Philosophia Verlag.

Thouless, D. (1989). Condensed matter physics in less than three dimensions. In *The New Physics*. Pp. 209-235. (Ed.) Davies, P. Cambridge: Cambridge University Press.

Vygotsky, L. S. (1934/2001). *Tänkande och Språk*. Göteborg: Diadolos.

Vygotsky, L. S.(1993). *Thought and Language*. Cambridge: MIT Press.

Wertsch, J. (1998). *Mind as action*. Cambridge: Harvard Univerty Press.

Woodworth, R. S. (1958). *Dynamics of Behavior*. New York: Holt.

Index

Abnormal mental states,. *Se* Superiority complex,
Abraham,
 Patriach prophet,. *Se* Jews
Accumulated myths, 158
Actualisation of potentialities, 49
Adaptation,
 process of, 155
Advancement,
 of knowledge, 158
Advertisements,. *Se* propaganda
Africa,
 black people of, 84
Aggressive behaviour, 159
Agreement approach, 102
Alfvén, Hannes,
 A Swedish physicist, 86
Alliance pattern, 75
Alliance structure, 75
Astronomy, 12
Auditory hallucinations, 159
Axioms, 64
Axioms,. *Se* Principles,
Barbarians, 17
barbarism, 157
Bias predisposition, 160
Blood language,
 utilisation of, 158
Book of Genesis,. *Se* Moses
Camouflage, 160
Canaan,
 Ham's son, 85
Carl Stumpf,
 German philosopher, 153
Carl von Linné,
 a Swedish Botanist, 154
central thesis, 153
Chance, 154
Chemical reaction theory,
 chemical theory, chemistry, 92
Chemistry, 158
Christians, 155
Churchill, Winston,
 British Prime Minister,. *Se* Second World War
Citizens of,. *Se* Western Europe
Classroom, 125
Climatic Law, 11
Cognitive constructivist perspective, 126
colonial era, 9
Colonial masters, 95
Common Safety Principle, 35
Competition, 155
Complex,. *Se* Superiority
Conditioning,
 Operant, Classical, 16

Consensus,
 definition,. *Se* Solidarity
contemporary lifestyle, 153
Contract,
 union, immigrants, 37
Criminals,. *Se* psycho s complex
Curriculum, 128
Darwin, Charles,
 Theory of evolution, 87
Darwin's hypotheses, 156
Darwinism, 94
Decontextualisation, 133
Deductive approach,. *Se* geometrical fashion
Definition, 14
Delusion,. *Se* symptom
Deprivation, 54
Development, 158
Discernment, 139
Discrimination, 10
Discrimination,
 origins,. *Se* colonial masters
Displacement,, 91
Dominant principle, 43
Dominant principle,
 sovereign, 41
Domination, 161
Drives, 15
Eastern Europe,
 democratic government, 51
Egalitarian alliance, 75
Environmental pressure, 161
Equilibrium, 49
Ernst Mach,
 philosopher, 153
Ethnomethodology, 125
Europe,
 Western countries, 13
Evolution,
 doctrine of, 156
Evolutionary theory, 157
Exaggerations, 161
Fascists,
 Mussolini, 156
Foreign organism,
 migrants,. *Se* Immigrants
Generation, 155
Genetic make-up, 161
geometrical fashion, 9
Geometry,
 branch of Mechanics. *Se*
German Marshal Helmuth von Moltke,
 comment on war, 156
Government, 10
Habituation hypothesis,

novel stimuli, 15
Ham,
    Noah's son, 85
Hammurabi,
    Code, law, 90
Ham's sons,
    Cush, Mizraim, Canaan, 86
Hannibal Odyssey,
    psycho s complex, criminal, aggression, 161
Henrich Himmler, 159
Heritable variation, 154
Hermann Helmholtz,
    German philosopher, 153
Hidden variables,
    motivations, 59
Hierarchic alliances, 75
Higher equilibrium, 50
Hobbes, Thomas,
    English distinguished philosopher, 36
Homo sapiens,
    All descendants of the human race, 92
Hostilities, 159
Human animals,. *Se* urmänniskor
Human Behaviour, 11
Hypothesis,
    discrimination,. *Se* Proposition
Identification,
    certain ideologies, 15
illusion,
    superiority as,. *Se* Superior species
Illusory superior power, 160
Imitation, 96
Immigratory Alliances, 76
imperialist regimes, 9
Imperialists,
    regime, 158
Impetus Principle, 30
Industrial capitalism, 156
Inert, 19
Infantile notions,. *Se* ideology
Inhabited universe, 159
Inheritance, 157
Irreversibility, 49
Jews,
    Hebrews, 39
Karl Marx,
    a philosopher, 156
Kinetic, 19
King Nebuchadnezzar,
    Babylon, 159
knowledge, 153
Language, 133
language,
    Original, blood, 11
Learning in the classroom, 125
Legal conflicts, 78
Living organism, 158
Low equilibrium, 50
Malthus,
    economist, 155

Manipulative,
    tatic, 160
Marriage Alliances, 76
Merits,. *Se* Quota
Messiah,
    Jesus Christ,. *Se* Jews
Migrated, 88
Minutrons Perforation, 69
Minutrons,
    charged mental ideas, wave, ether, 18
miserable condition, 153
Mobbing,
    immigrants,. *Se* migrants
modern technology, 153
Moses,
    Author of Genesis, 85
    Ten commandments, 90
Multiple Alliances, 77
Narcissistic,
    psychiatric symptom, 160
Natural selection,. *Se* Charles Darwin
Nebuchadnezzar,
    Babylonian King, 54
Newtonianism, 154
Noah,
    biblical mythology of, 84
Non-discrimination law, 10
Nostrademus,
    dreams, prophesy, 160
Observation, 158
Observation approach, 102
Obsessed, 159
Omega Theorem, 31
Omnipotent wishes,. *Se* infantile notions
Organism, 14
Partner,. *Se* Union, contract
Pavlov, Ivan,
    Russian psychologist,. *Se* Conditioning
Peripheral legitimate participation, 132
Phenomenography, 125
Phenomenography and variations perspective, 126
Phenomenology, 125
Phenomenon,. *Se* Discrimination,
Physics, 12
Physics,
    Physical science, the universe,. *Se* Astronomy
Pinnacleconductivity, 55
Political Discriminology, 12
Postulate, 18
Predators, 154
Predictive power,
    superiority complex theory, 162
Principles, 11
Principles of Geology,
    C. Lyell's book, 154
Principles,. *Se* Axioms,
Professional Alliances, 75
Progeny, 154
Propaganda tactics, 161
Propaganda,

Adolf Hitler, 160
Proposition, 61
Prudence, 103
Psychiatric symptoms,. *Se* psycho s complex
Psychological laboratories, 125
psychological superiority complex, 153
Psychophysics, 158
Psycho-S Complex,
  signs, symptoms, 159
Puppet sovereign,
  among the natives, 44
Purchase approach, 102
Queen of Sheba,
  Ethiopian Queen, Ancient world, 39
Racism,. *Se* Discrimination
Racist-militarist,
  Hitler,, 156
Recognition, 49
reductionism that took place in, 9
religion,
  blinded men, 9
  repercussions, 153
Roboantropos, 70
roman conquerors, 9
Saamis,
  The original natives in the land of sweden, 47
Santa Barbra Research Group, 135
Sapience, 99
Saul of Tarsus, 159
Saul,
  King of Israel, 159
Savages, 17
Scholium,
  Latin, comments,. *Se*
Second World War, 68
Selfishness, 60
Selfishness,
  theory of, 49
Self-perception, 49
Self-sufficiency, 49
Sensorimotor, 127
Sexual selection, 157
Sigmund Freud,
  psychiatrist, neurologist, 153
Simultaneity, 139
Situated learning theory, 132
Skinner, B. F.,
  American psychologist, 29
Slavery,. *Se*
Social categories. *Se* Darwin's theory
Social Darwinian, 156
Society,
  a model for, 156
Socio-constructionist perspective, 128
Sociocultural and the community focused
  perspective, 126
Sociocultural and the Interactionistic Perspective,
  129
Sociocultural perspective, 132
Sociolinguistics, 125

Socrates,
  Greek philosopher, 39
Solidarity,
  politics as, 61
Solomon,
  Hebrew King, 39
Sovereign Alliances, 76
Sovereign status,. *Se* Dominant principle
Species,
  theory of, 154
Speculative thinkers, 154
Stealing approach, 102
Strictus, 101
Subjectuses,
  Immigrants with citizenship. *Se*
Superexaggerare,. *Se* exagerations
Superfluity, 54
Superior species, 9, 95, 161
Superiority complex, 49
Superiority,
  theory of, 10
Superordinate, 29
Superwunsken,
  portentous wishes, 161
Syllogism, 105
technological ages, 153
The Classroom Discourse Group, 135
The Cognitive Constructivist Approach, 127
The Descent of Man,
  Darwin's theory, 155
The enlightenment era, 158
The Galapagos,
  an Isaland, 157
The genius approach, 102
The Hard Conditioning Incentive,. *Se*
the Jews,
  Nobel Prize, 159
The Language Focused Approach, 135
The Lucifer Theorem, 32
The Main *Frontis*, 98
The Origin of Species,
  Darwin's book, 154
The Primary *Frontis*, 97
The secondary *frontis*, 98
The secondary law of intelligence, 104
The Soft Conditioning Incentive,. *Se* Government
The spies approach, 102
Theory of Alliances,. *Se* Discrimination,
theory of superior species, 154
Underground Alliances, 76
Uniformatarianism,
  theory of geology, 87
uniformitarianism,
  theory of, 154
Universe, 29
Urmänniskor
  a Swedish word, 159
Variables, 15
Variation, 139
William James,

physiologist, philosopher, psychologist, 153
Winmust, 160

Voluntarius Alliances, 77
World War II, 47

Author's summary

To develop a new scientific theory that explains a phenomenon is not an easy enterprise, especially when it concerns social behaviour. Few have this gift where they propound new theories about nature, human behaviour, or the physical world.

Emulating well-known theorists in the twentieth century and a few other geniuses, this author theorises on discrimination behaviour in work places using the deductive method, and offers significant catching propositions that need to be investigated and as a result set the reader into deep thinking.

This treatise offers a number of sounding propositions such as "War is a symptom of a disorder," "Superiority is attained not given," and "We are all Homo Sapiens," in the attempt to champion the banishment of this negative behaviour in this modern society.

Desmond Ayim-Aboagye received his B.A., M.A. from Andrews University (USA). Later he studied at the Uppsala University (Sweden) in the field of Psychology and Religion and Educational Sciences, and then received M.A. PhD in Psychology of Religion in 1993. He became Associate Professor in 1997 at the Åbo Akademi University, Finland. Over the past decades he has had clinical experiences in the Psychiatric Clinics in Sweden, such as Uppsala University Psychiatric Hospital and Huddinge University Psychiatric Hospital in Stockholm. He has taught and lectured in different universities such as Uppsala, Örebro and Mäladalen Universities in Sweden, and University of Ghana, Legon. Currently, an active regional politician in Sweden and a Senior Lecturer at the Uppsala University besides his scholarly work, he has published articles and many books, including *The Function of Myth in Akan Healing Experience: A Psychological Inquiry into Two Akan Healing Communities (1993); The Psychology of Akan Religious Healing 1997); Using Christian Religious Resources in the Welfare of Prisoners: The Case of Swedish Prisons (1997); Prison, Punishment and the Church. A Socio-Psychological Investigation of the Work of Chaplains among the Immigrant Inmates in Swedish Prisons (2000); Punishment, Prison and the Clergy. A Social Psychological Perspective (2006); Indigenous Psychiatry. Transcultural Study of Traditional Practitioners in West African Healing Communities with Focus on Ghana (2006); The Psychiatric Care in West African Mental Hospitals: The Impact of Religion and Tradition on the Care of Mental Patients* (A book in progress)